Up to Speed

Up to Speed

THE GROUNDBREAKING SCIENCE OF WOMEN ATHLETES

Christine Yu

Riverhead Books
New York
2023

RIVERHEAD BOOKS
An imprint of Penguin Random House LLC
penguinrandomhouse.com

Small portions of chapter 4 appeared, in different form, in the
author's article "The Condition That's Quietly Sidelining Female
Athletes" in *Outside* (September 15, 2017).

Library of Congress Cataloging-in-Publication Data
Names: Yu, Christine, 1976– author.
Title: Up to speed : the groundbreaking science of women athletes /
Christine Yu.
Description: New York, N.Y. : Riverhead Books, 2023. | Includes
bibliographical references and index. |
Identifiers: LCCN 2022033690 (print) | LCCN 2022033691 (ebook) |
ISBN 9780593332399 (hardcover) | ISBN 9780593332412 (ebook)
Subjects: LCSH: Physical fitness for women. | Women athletes—
Physiology. | Sports for women. | Sports sciences.
Classification: LCC GV482 .Y8 2023 (print) | LCC GV482 (ebook) |
DDC 796.082—dc23/eng/20221006
LC record available at https://lccn.loc.gov/2022033690
LC ebook record available at https://lccn.loc.gov/2022033691

Printed in the United States of America
1st Printing

BOOK DESIGN BY AMANDA DEWEY

To Dad:

For showing me science is cool and teaching me
to be curious about the human body.
Miss you always.

To Mom:

For everything.

Her wings are cut, and then she is blamed for not
knowing how to fly.

—Simone de Beauvoir

Everyone thinks women should be thrilled when we get
crumbs, and I want women to have the cake,
the icing, and the cherry on top, too.

—Billie Jean King,
thirty-nine-time Grand Slam tennis champion

CONTENTS

• • ● ● ●● • •

INTRODUCTION
Mind the Gap

It happened on my last 400-meter repeat on the track. As I came into the backstretch, I heard a visceral, gut-twisting sound—a pop and click in my right knee. It was like I cracked the joint, the noise similar to cracking my fingers, except I didn't feel the same release and relief. Within a step or two, I knew I couldn't bear weight on my leg; when I did, it felt like my knee would crumble beneath me. I hopped and limped—anything to slow my momentum and bring my body to a stop—and muttered, "Crap." I made my way across the infield to a bench and called my husband to pick me up. My goal of running a half-marathon personal best was very likely off the table.

That was February 2012. Fifteen years earlier, I tore my ACL—the anterior cruciate ligament—and medial meniscus in the same knee while studying abroad in Italy. A group of friends and I took a weekend trip to ski in the Swiss Alps, a dream of mine. But a random, weird fall resulted in a ride down the mountain tucked into a ski patrol sled. It was a comedy of errors. I couldn't communicate with the ski patrolman. He spoke German and I spoke English and some elementary Italian. In the end, he waved his hands at me, flustered. He pulled out a cube of sugar, doused it in brandy, stuck it in my mouth, and strapped me in the sled. My friends followed behind, giggling and snapping pictures.

The reality of the situation didn't set in until the adrenaline wore off, and I broke down in tears in the medical office at the base of the mountain. I had surgery a few months later when I returned home to California. It was a long, frustrating road back to all the sports and activities I loved, a journey I never wanted to repeat. My knee is still always in the back of my mind, whether I'm running, weight lifting, surfing, and especially skiing. I'm constantly worried about hurting it again. With every ache or pain, real or phantom, I'm afraid I've injured something.

So, when I heard that pop and click on that winter morning, my mind immediately started running through decision trees and planning for the worst. In the days and weeks that followed, a mixture of anger and frustration boiled in my stomach. I didn't understand what had happened. Who tears their ACL running on the track? I was embarrassed and ashamed that I was injured—again—and I couldn't stop thinking there was something wrong with my body. That I was prone to injury. That it was written somewhere in my DNA that I wasn't cut out for certain activities and sports.

I thought I knew a fair amount about the human body, exercise, and fitness. I come from a family of doctors—my dad, my sister, and countless cousins all practice medicine—and I was destined for the same path after college. I believed in science. I trusted it. I did everything by the book to rehabilitate my knee the first time around. I diligently went to physical therapy, did all my prescribed exercises (no matter how painful or boring), and rebuilt my strength. Leading into this half-marathon training cycle, I regularly lifted weights and gradually increased my running mileage. Yet following the rules didn't seem to matter. Even my orthopedic surgeon had no answers. I remember him saying, "Christine, I don't know what to do about you." Two months later, I had surgery to replace my ACL again.

I'm not the only one whose body has had a fraught relationship with sports and exercise. As a journalist, I've noticed that even as women have excelled in sports, there's an underlying sense that women and their bodies are an anomaly in the athletic world. Women athletes are often led to believe that menstrual dysfunction is normal, stress fractures are a rite of passage, knee injuries are inevitable, disordered eating and body image issues are part and parcel of the athletic experience, and athletic careers are limited by puberty, pregnancy, and age. When something goes wrong—injury, burnout, overtraining—women are blamed and shamed for it, despite doing everything they're "supposed" to do. But I couldn't figure out why the feeling was so pervasive and intractable. Why weren't we taking better care of girls and women? Did it have to be like this?

The pieces started to click together during the summer of 2018. At the time, I was reporting on an article about the field of sports science and I was trying to understand why so much scientific research leaves out women. The bias against women in biomedical research was something I knew about on a subconscious level, but I admit, I never really thought about it explicitly. As a journalist who covers sports, science, and health, I regularly come across studies that involved men as participants. When I couldn't find research that included participants who were women, I gave scientists the benefit of the doubt. I assumed the lack of representation wasn't intentional. There had to be a perfectly reasonable explanation. Maybe women were harder to recruit. Maybe women weren't interested in participating in studies.

Then I fell down a deep rabbit hole.

As I read more papers and talked to experts, I began to understand that the exclusion of women was more than an oversight. It became clear how, for decades, scientists have worked under the assumption that women and men are biologically and physiologically the same, if you just ignore the reproductive organs. How so much of what we take as gospel about exercise training, nutrition, performance, and injury prevention is based on what's found in human participants who are men or experiments with male cells or animals. How scientists don't know for sure whether those recommendations apply to women. How they didn't see a need to address those data gaps.

But "women are not small men," as Stacy Sims, an exercise physiologist and nutrition scientist who studies sex, gender, and sports science, told me. Yet that simple proposition underpins the current system of sports and science. And ignoring important biological, anatomical, and physiological differences between women and men can have real, negative implications outside the laboratory.

It's a curious disconnect. The number of girls and women taking part in physical activity, from fitness classes and recreational leagues to professional sports, has risen dramatically over the last fifty years. In the United States, 3.4 million girls play high school sports. At the collegiate level, the number of women playing sports has increased 600 percent from the early 1970s. This groundswell is a global phenomenon. At the Tokyo Summer Olympics in 2021, women made up nearly 49 percent of the competitors, making them the most gender-equal Games in Olympic history.

But the fields of exercise and sports science haven't kept pace with this rising population of active women. Even in the twenty-first century, coaches, doctors, trainers, and athletes themselves know the bare minimum about women's health, and women struggle to find care and advice to help them feel their best. There isn't much research that looks at the factors that influence athletic performance or the effectiveness of different training and nutrition interventions among women.

There's another inherent, deeper contradiction embedded in the fabric of science and of sports. In the scientific literature and research laboratories, the features that make a body distinctly female aren't important enough to warrant additional investigation, but in the world of sports, that viewpoint is flipped on its head. The very characteristics that scientists cast aside as no big deal—the menstrual cycle, hormones, the uterus, ovaries, breasts—are the reason women were marginalized and presumed to be unfit for sports, or at least in need of a segregated competition category. The confluence of these two beliefs has had a huge impact on the athletic lives of women.

When I first started reporting this book, I thought, somewhat naïvely, that this would be a story about the body—hormones, bones, muscles, aerobic capacity, ligaments, tendons, and the myriad ways women's bodies are not the same as men's. I wondered if those differences mean we need to toss everything we know about sports science out the window and start over from scratch. Do women need a completely different approach to training, nutrition, injury prevention, well-being, and health?

I quickly realized it wasn't so simple. We can't just put men in one basket and women in another and treat them as two distinct organisms. There's a whole messy middle where the groups intersect physiologically and biologically like the circles of a Venn diagram. Right now, we don't know the nature of the overlap of those circles—how big it is and what it entails—nor do we know a lot about the part of the women's circle that stands on its own. We've only just begun to interrogate the differences and similarities that exist between women and men and how significant they may be in the grand scheme of things.

When women and men aren't included in research studies in equal numbers, we potentially miss important insights pointing to the diverse routes people may need to take to reach the same goals, whether it's being a lifelong athlete, racing well, or staying injury-free. Without a better understanding of all humans, we don't know how to guide people along those paths.

This lack of knowledge and awareness is what makes it hard for women, especially active and performance-driven women, to find the resources they need, even when they ask for help. Pro runner Leah Falland spent years looking for answers. She had a stellar collegiate career at Michigan State University. She won two individual National Collegiate Athletic Association (NCAA) titles, helped Michigan State to a national team title in cross-country, and developed into one of the country's top competitors in the steeplechase, an event that involves hurdling over twenty-eight fixed barriers and seven water jumps over 3,000 meters of track. She told me she was supposed to be "one of the greats," racing all over the globe as a world-class athlete.

But Falland's body never seemed to cooperate with her ambitions as a professional runner. She sustained a litany of injuries—a ruptured plantar fascia; stress fractures in her foot, shin, and pubic bone; a shredded labrum in her hip. Each injury was a threat to her career, and Falland became sad and anxious. She was exhausted down to her bones. Her hair fell out. Her period was irregular. She gained over fifteen pounds. She was grumpy and didn't want to date or be social with other people. Unbeknownst to her, she was anemic and, after years of undereating and overtraining, her thyroid wasn't working properly.

Outside of diagnosing and treating her acute injuries, doctors couldn't figure out what was wrong because no one knew how to look at her holistically—as an athlete, a woman, and a human. General practitioners don't always understand the nuanced context needed to treat elite athletes and interpret their diagnostic tests and lab results. Since Falland never had her blood tested in college, doctors didn't have a baseline to compare her numbers; they only had what was considered "normal" for adult women. Those ranges aren't always meaningful if you're active because they're based on the general population; they can mask subclinical abnormalities that can influence fitness. For instance, if a woman athlete's bloodwork shows that ferritin, an indirect measure of iron, is less than 20 µg/L, her doctor may not flag it as a concern. Technically, it's within the healthy range, albeit at the lower end of the spectrum. But since iron directly affects the body's ability to transport oxygen, athletes may require higher levels for optimal performance. Low ferritin levels may be indicative of iron deficiency without anemia and contribute to symptoms like fatigue and a weakened immune system.

For Falland, doctors assumed she was healthy because her markers appeared normal. Another specialist suggested that she just needed some supplements and an IV of vitamin B_{12}. Other times, as soon as doctors heard that she experienced anxiety and depression, they wrote off her symptoms as side effects of stress. If she just meditated and got more sleep, everything would be fine.

In the course of one year, Falland spent $15,000 bouncing from doctor to doctor. "It got harder and harder to ask for help and I got pretty worn down," she told me. Every avenue led to a dead end, and it took a toll on her. At age twenty-six, three years after her struggles began and with a new stress fracture in her femur, she was ready to walk away from the sport she loved.

Without better information, talented athletes like Falland may leave their sport behind because they haven't been looked after properly and can't train to the best of their ability. "Finding a good doctor, especially as a woman, can be very difficult," Falland says. Even now, she's still a little on edge when she sees a physician; she's afraid they're going to ignore her or they won't believe her. Eventually, she found a skilled team that understands women's bodies, particularly athletic ones, and they helped her get healthy and back to elite running. At twenty-nine, she joined On Athletics Club, a professional running team coached by three-time Olympian Dathan Ritzenhein, and is running well again.

When Falland looks back, she mourns the years she lost in her early twenties because the medical system as well as the system around sports didn't understand her as a woman. "It was supposed to be the best time in my career, but I feel like I lost those years and parts of my young adult life," she says.

The marginalization and exclusion of women isn't strictly a sports story or a science story, but sports and science are both predicated on seemingly straightforward conceits. Sports is about competition and performance, each athlete striving to get the best out of themselves. Science is an iterative process that is assumed to be neutral because it focuses on data and facts. It's also an incremental process, and institutions that fund scientific research want to see that new projects build off an existing evidence base.

Yet sex and gender bias destabilize what's supposed to be a level playing field. Who gets excluded matters. Jan Todd, a noted sports historian, professor, and world-record-setting powerlifter who pioneered the sport for women, told me that because women were left out of scientific research and the historical narrative of sports from the beginning, they continue to be overlooked—there's no precedent to include them. This act of omission creates a blind spot, enabling society and institutions to routinely brush aside women and position them as the exception to the rule rather than part of the norm.

So, why aren't we studying women?

In the 1960s, the field of sports science began to change on the heels of disappointing performances by Americans at the Olympic Games and the emergence of the Soviet Union as an athletic power-house. Americans had consistently topped the medal count before, but now they were behind. After making its debut at the 1952 Summer Games in Helsinki, Finland, the Soviet Union outstripped the Americans by roughly thirty medals at the next two Olympics.

One reason the Soviets triumphed was their systematic sports science program, designed to improve athletic performance. While some of the better results are believed to have been due to the use of anabolic steroids, the program was still light-years ahead of anything in the United States. In America, coaching and physical education were more art than science. Coaches gathered intel from military fitness and medical rehabilitation articles and then made their best guess as to how those protocols applied to athletes. In 1963, James B. Conant, former president of Harvard University, published a blistering criticism of the education system in his book *The Education of American Teachers*. In particular, he was "far from impressed by what I have heard and read about graduate work in the field of physical education." In his opinion, the discipline lacked academic rigor, and he recommended canceling all graduate programs in the field.

Conant's report sent physical educators and university administrators scrambling to prove the discipline was a scholarly field, one that produced and disseminated research, not just future teachers and coaches. Subdisciplinary research specialties—sports history, biomechanics, sports psychology, and exercise physiology—began to emerge as the field shifted from "physical education" to "kinesiology." "They weren't only training teachers anymore. They were trying to advance sports science," says Carole Oglesby, a kinesiology professor and trailblazing activist for women's sports.

Jan Todd told me how this played out at the University of Texas (UT), where she is a professor. Up until 1973, UT had three different programs. There was women's physical training and men's physical training, which taught the required physical education classes for students. The programs were run separately with separate gymnasiums for women and men. The third program was physical education. Housed in the College of Education, it was designed to prepare future physical education teachers.

In 1973, the three programs merged, and Todd says the constitution of the department shifted pretty quickly. "Being somebody who just teaches physical activities doesn't really matter in the same way that it did before," she says. The number of physical activity classes declined, and the department hired more scientists and researchers, mostly men, who started to bring in funding to support their work. Soon, "they became the most important people in the department," Todd says.

This shift happened at the same time as the passage of Title IX of the Education Amendments of 1972, which prohibited sex-based discrimination by any education program or activity receiving federal funds. To comply with the new law, physical education programs tried to shoehorn women into athletic departments that typically housed men's programs and sports. Oglesby remembers that women called these mergers "sub-mergers" because "the men's programs were the superstructure on top of the women's programs," complete with its greater emphasis on elite sports, scientific research, and men. In other words, the consolidation solidified a system where sports and sports science centered on men.

Even now, men's voices continue to dominate research agendas, funding priorities, and resources within the field. Male anatomy, physiology, and biology inform standards for athletic development and progression, guidelines for training and nutrition, and blueprints for athletic shoe, clothing, and gear design. These benchmarks weren't constructed with girls and women in mind, yet girls and women have always been assessed against these norms. When women don't conform to the accepted mold, it's no surprise that their so-called shortcomings are blamed on their bodies, especially features that distinguish women from men—chromosomes, hormones, genitalia, gonads, and secondary sex traits. These characteristics are then labeled as deficits and defects because they don't match the male model.

It's easy to internalize the idea that there's something wrong with your body rather than with the systems that didn't consider you in the first place. Consciously and unconsciously, girls and women are conditioned to ignore and dismiss the traits that make their bodies uniquely female, especially in the context of exercise and sports. They don't talk about periods or cramps. They try to eliminate or control menstrual cycles. They smush their breasts down as much as possible. They hide the fact that sometimes they pee a little bit when they do jumping jacks or box jumps.

But differences in biological sex aren't the only reason women are marginalized. Gender plays a role too: the socially constructed identities and roles that determine how people perceive and present themselves and relate to others. These identities are informed by prevailing sociocultural values, attitudes, and politics, and sports reinforces these beliefs. Sports has always been a gendered space, an arena that privileges men and allows them to test and display their masculinity. When the athletic world is set up to prioritize the perspective of men, gender becomes another means of exclusion for anyone who doesn't identify with the predominant view.

Throughout much of history, women, widely considered physically inferior to men, have been kept off the pitch, the court, and the starting line. Physical exertion and competition were seen as incompatible with a woman's role in the home, her fertility, and her femininity. In the late nineteenth and early twentieth centuries, when women began to participate more in exercise and organized sports, it was still a realm where men flourished. At the 1900 Olympics in Paris—the first Olympic Games in which women were allowed to participate—only 22 of the 997 competing athletes were women. More recently, even as women have defied long-held beliefs about their athletic capabilities, men continue to occupy center stage and women are treated like footnotes to the main story.

I n the fall of 2019, the ground shifted in the way we think about women and sports, and conversations about women athletes rose to the forefront in ways they haven't in the past. *The New York Times* published a video op-ed by Mary Cain that illustrated the very real consequences of living in the shadows of men in sports. In the early 2010s, Cain was a high school track phenom, considered a once-in-a-generation athlete and the future of American middle-distance running. She said she wanted to be the "best female athlete ever." Full stop. At age seventeen, she was offered the chance to pursue her dreams alongside Alberto Salazar, then considered the best coach in track and field. It was as if Vince Lombardi called up a high school quarterback and asked him to play for the Green Bay Packers. Cain packed her bags and moved from her family's home in Bronxville, New York, to Portland, Oregon, to train with the Nike Oregon Project (NOP).

But her dream quickly turned into a nightmare. Cain was still in adolescence, a period of tremendous growth and development, but the athletic environment she was placed in wasn't set up to nurture her changing body. Instead, she was expected to fit the developmental timeline and athletic milestones more typical for boys and men, not a teenage girl. Her body deteriorated: She developed an eating disorder and missed her period for three years. She broke five bones as a result of malnourishment and was at risk of osteoporosis. She harmed herself physically and began to consider taking her own life. Unsurprisingly, her performances tanked. She quit the team, leaving her dreams of competing in the Olympics behind her. (Right before Cain's revelations, Nike shut down NOP due to unrelated doping allegations involving

Salazar, for which he is serving a four-year coaching ban issued by the U.S. Anti-Doping Agency. Salazar has denied any wrongdoing in both the doping case and mistreating Cain.)*

Cain's op-ed reverberated both inside and outside the running world. It was everywhere on my social media feeds, not because her story was exceptional but because her experience is so common. Cain spoke candidly about what it's like to be a woman in elite sports, centering her experience in conversations that have traditionally expected women to paint a glossy, feel-good picture. In doing so, she made it okay for other girls and women to talk about their own stories—the good and the bad. She made it okay for others to question the pervasive myths about taboo topics like eating disorders and missed periods. She made it okay for them to reckon with their own at times fraught experiences because it wasn't their fault. The athletic and sports science system was designed by and for men; it wasn't developed to promote the health and performance of women athletes because it ignores their specific experience in sports.

Cain's story was pivotal for another reason. Conversations that had begun to percolate within sports and exercise science circles about the underrepresentation of women in research studies, like the one I had with Stacy Sims, began to reach larger audiences outside of laboratories. People became interested in the science and wanted to peel back the layers of assumptions behind the training and nutrition guidelines that women have long relied on. They were invested in understanding how those principles relate to women's experiences of exercise and sports.

In the years since Cain's story became public, social media has been awash with similar accounts from women and girls across the world and different sports—from current and former pro athletes to amateur and age-group competitors—unleashing the rallying cry to #FixGirlsSports.

---

* The U.S. Center for SafeSport, an independent organization that investigates physical, sexual, and emotional abuse in Olympic and Paralympic sports, also issued Salazar a lifetime ban from the sport for an alleged sexual assault.

What's become clear is the current system of sports and science isn't built to accommodate diverse points of view and doesn't meet the needs of all people who wish to be active. Our understanding of exercise science, sports nutrition, and injury prevention and treatment is based on a model of men—one that treats male physiology and men's developmental curve as the gold standard and doesn't always investigate the important biological, anatomical, and physiological differences between women and men. Cain's experience demonstrated what can happen when women fight their bodies in order to fit into protocols based on a partial picture of the human population.

The lack of high-quality research on women means there are no evidence-based sports guidelines tailored to women—whether they're elite or recreational athletes. The repercussions for elite athletes may mean sacrificing medals, championships, podium finishes, the opportunity to compete, and even long-term health. For recreational athletes and active girls and women, it creates barriers to getting into and staying involved with physical activity and sports. It may mean injury screening and prevention protocols that miss the mark, leaving women vulnerable to head and musculoskeletal injuries. It may mean uncomfortable gear and sports bras that don't fit the proportions of women's bodies. It may mean that girls and women sell themselves short because their fitness and athletic progression doesn't look like it "should."

Anthony C. Hackney, a professor of exercise physiology and nutrition at the University of North Carolina at Chapel Hill, first started studying women athletes as a graduate student back in 1979. He told me that in the past, it was challenging to rally support behind researching women, leaving the field understudied and poorly understood. "Here we are, forty years later, and I'm wondering, Why is half the population of the world being ignored within certain facets of science?" he says. While he's glad to see a new sense of awakening and interest building around women athletes, he's still a little puzzled. He asks, "What took everyone so long?"

I wrote this book, in part, to answer Hackney's question. Why are women excluded from exercise physiology and sports science research? What are the implications of this data gap for girls and women? How can we empower people to excel, not despite (or in spite of) their sex and gender but because of it? And yes, what took everyone so long? People are asking these questions—on the court and pitch, on the trails and track, at the pool and gym, in universities and labs across the world.

This book starts with the story of how sex and gender bias came to corrupt the systems of scientific research and sports and why men are privy to so much more information about their bodies than women. It reveals how discrimination against women has impacted their role in sports—their participation, training, performance, injury rates, and long-term health and well-being.

From there, the book dives into specific topics, including the menstrual cycle, nutrition, endurance, injury, breast health, and gear. By examining the latest research and stories of athletes themselves, each chapter disentangles the interplay between myth, gender bias, science, and real lived experiences. It shows how researchers, athletes, and others are pushing back against the status quo and proposing new solutions to better serve the needs of women and girls.

Finally, the book looks at the challenges women face during three critical life stages: adolescence, pregnancy and the postpartum period, and the menopause transition. These final chapters provide a road map to help women navigate the unique concerns and distinct priorities they face during these phases. They also address what we could do on an individual, cultural, and institutional level to create a better, more inclusive system of sports and science.

Talking about humans and their bodies within the context of sports and science is a tricky endeavor. Both disciplines rely heavily on binary categories that often conflate sex and gender as one and the same. Organized sports are based on sex-segregated competition categories, either women's or men's. Scientific research builds an evidence base through controlled models and experiments. To eliminate variation and keep things simple, specimens and participants are classified as either female or male. Yet these either-or definitions are static and don't accurately describe the larger spectrum of sex or gender. They unwittingly reinforce stereotypical notions of women and men, femininity and masculinity. Cisgender people, those whose sex assigned at birth aligns with their gender identity, are more likely to fit into these categories. For transgender, nonbinary, and intersex people, these categories don't always adequately reflect their identities, yet they are forced to pick one or be left out.

Throughout the book, I use the terms "female" and "male" when they pertain to sex-related features of anatomy, physiology, and biology and not as a way to describe or identify any specific person. When the science pertains to someone assigned female at birth, I use more inclusive language where possible, such as people with breasts or people who menstruate, recognizing that not everyone who has breasts or menstruates identifies as a woman. While we're most accustomed to using terms like "girl/woman" and "boy/man," there's a tremendous range of diverse gender identities, which may or may not align with the sex a person is assigned at birth. Generally, I refer to "girls," "women," "boys," and "men" when the term reflects a person's identity, the historical context, or a socioculturally constructed context like women's sports or women's health. The usage of these terms isn't meant to be exclusionary. I know it's not perfect, but I hope it can help us think more expansively about sex and gender.

Ultimately, the systemic injustices and assumptions embedded in the fields of sports and science don't just affect women and girls. They affect all people. By examining how women and girls are marginalized and excluded, we can start to uncover ways to make these systems more equitable for everyone. This opens a whole new world of possibilities and better information to help more people get involved and stay involved in sports and physical activity.

Here's the exciting part. Even within the male-centric model, women have accomplished incredible feats of athleticism, smashing long-held beliefs about what women can or cannot achieve. Imagine what would happen if everyone had access to better information and guidance.

Up to Speed

1.

Where Are All the Women?

• ◦ ● ◉ ● ◉ ● ◦ •

In 2016, mountain biker Kate Courtney neared the end of her under-23 career and was ready to make the leap to the next level. She started working with coach Jim Miller, then the director of high performance at USA Cycling and now the organization's chief of sport performance. Miller knows a thing or two about what it takes to develop Olympians and world champions. In the last two decades, his athletes have won fourteen Olympic medals and multiple world championship titles. Together, they built a plan not only to prep Courtney for what was supposed to be the 2020 Summer Olympics in Tokyo but to set her up for a long-term career at the next stage of elite competition.

To succeed in cross-country mountain biking, Courtney needs different tools in her arsenal—explosiveness to get off the starting line quickly; technical skills to navigate tight, single-track turns and through rock gardens; power to churn up steep climbs; vision to spot fast lines; mental calm to handle unexpected situations; and endurance to gut out a ninety-minute race that she describes as "sprinting a marathon." There are a number of levers Courtney and her team can play with to optimize her performance potential. Given her laser focus on being the best athlete, it's not surprising to learn that Courtney's a data nerd who wants to understand how the human body works. (She was a human biology major at Stanford University, after all.) It suits her sport, where there is no shortage of numbers, statistics, and trends to mull over and analyze. She has worked meticulously to dial in her nutrition and training—lung-searing intervals on picturesque trails across the Golden State, gym sessions that involve a mind-boggling mix of box jumps and Olympic lifts, and yoga for recovery—all of which has helped her climb the competitive ranks, both in the United States and internationally.

But the world champion and Olympian isn't one to blindly go by the numbers or apply prescriptive exercise and nutrition guidelines. She looks at research studies to understand the why behind the recommendations and determine if they make sense for her as an athlete and a person. "I'm always trying to look at the underlying picture and make sure I'm applying the right thing to my training and to myself individually," she told me. Except that when she started looking into the sports science studies, she quickly realized that the vast majority were based on men, and it wasn't entirely clear whether the findings held true for women too.

The process of conducting scientific research is like putting together a jigsaw puzzle. You work with thousands of disparate data points that, at first glance, don't seem to fit together. Just like you'd sort puzzle pieces by edge, color, and pattern, you start with the most basic research questions and work on those in small batches before moving on to thornier questions. As different pieces begin to match up, the accumulated data points resolve to form a clearer picture of what's going on.

In sports science research, scientists have studied men for more than a hundred years. It's like they've completed multiple sections of the puzzle—cardiovascular response, muscle adaptation, endurance, and biomechanics—and have started fitting them into a larger framework. We have a pretty good idea of what exercise means for active men and their bodies. Courtney's male mountain bike peers are working with a base of research that offers them more nuance to tailor their training and nutrition protocols to their physiology. It gives them a leg up on improving their health, well-being, and athletic performance.

Scientists didn't begin to study women athletes in earnest until the 1980s and 1990s. It's like they've just dumped out all the puzzle pieces and assembled the border, without a photo of the finished puzzle to guide them. Once all the pieces are connected, will it be the same picture as the men's puzzle or something else entirely? Courtney realized that the advice she'd read in books or received from others was based on either mixed evidence or one or two small studies on women. For a world-class athlete, that's a pretty rickety foundation to rely on.

Courtney shared her concerns on Instagram, writing, "Throughout my career, it has often been hard to see myself in sports science research. In many cases, women aren't included in the data at all or treated as a poorly understood 'special case.' What does it look like to train sustainability, fuel well, pursue longevity and stay healthy, happy and strong as a female athlete?" Right now, we don't have the answers.

Sex and gender bias have long been baked into the DNA of both athletics and exercise science. Modern sports developed in the late nineteenth century at a time fraught with change. The Industrial Revolution and urbanization ushered in massive disruptions to the way people lived and worked. People moved from the country to over-crowded cities, where gambling and drinking were common. Jobs required less hard labor, spurring concerns that men would become physically weak and "effeminate," Jaime Schultz, a professor who studies the history and culture of sports, told me. And weakness was perceived as a sign of a lack of moral character.

Sports were seen as an antidote to social ills and the so-called crisis of masculinity. It was a way to teach boys and men the virtues they needed to succeed: courage, grit, and self-control. During his presidency, Teddy Roosevelt praised "the strenuous life," particularly vigorous athletics like football. In the absence of war, men could test their mettle on the field. Players lined up in regimented, battle-like formations and used brute force to overtake their opponent's territory. The game purportedly created warrior-athletes and leaders—fit, muscular men of good character.

The idea of sports and exercise as a masculine domain persisted well into the twentieth century, particularly during the Cold War. Politicians were concerned that American youth were becoming less physically fit. By midcentury, the Selective Service rejected one out of every two young Americans because they were deemed unfit to serve in the military. Researchers found that nearly 58 percent of youth failed tests of muscular strength compared to only 8.7 percent of European youth. Losses to the Soviets in the 1956 and 1960 Olympic Games further stoked fears. In December 1960, president-elect John F. Kennedy wrote a piece for *Sports Illustrated* titled "The Soft American." To Kennedy, lack of physical fitness was a threat to national security. He wrote that "such softness on the part of individual citizens can help to strip and destroy the vitality of a nation." Americans—meaning American men— needed to get in shape.

When scientists became interested in studying human physiology and exercise, they naturally turned to able-bodied men as subjects. One of the most prominent research laboratories was the Harvard Fatigue Laboratory. When it was founded in 1927, the lab's mandate wasn't to study sports science per se. It was to determine the causes of fatigue so companies could improve the efficiency of factory workers. The scientists, unofficially led by physiologist and Harvard professor

David Bruce Dill, wanted to understand why the body tires, how it responds and adapts to stress, and how it maintains a "steady state." What better way to study fatigue than exercise?

To get the most out of their studies, the scientists sought out physically fit subjects, ones who could sustain high levels of work and repeat the trials over a period of days or months. Instinctively, they turned to athletes, including seven-time Boston Marathon champion Clarence DeMar, along with lab staff, many of whom were athletes themselves. In a couple of rooms in the basement of Harvard Business School, men (and sometimes dogs) ran or walked on a treadmill or rode a stationary bike (not the dogs).

The lab wasn't the only one to recruit young, active men for research purposes. Archibald Vivian (A. V.) Hill, a giant in the field of exercise physiology, long favored enlisting athletes in his studies because they could function at the upper limits of human capacity. For instance, while he was a visiting professor at Cornell University, he studied sprinters to understand how bodies recovered after a burst of high-intensity activity. Others at Yale University conducted experiments with the 1924 United States Olympic rowing team. After Harvard's lab shuttered its doors in 1947, its former students, staff, and fellows dispersed and went on to establish seventeen new labs across the United States, bringing the legacy and methodologies of their mentors with them.

Even today, it's often easier to find men to volunteer for exercise science research. There's a larger pool of men who likely have the caliber of fitness researchers are looking for. At the collegiate level, men make up 56 percent of the total student-athlete population in the NCAA. Men enjoy more than three times as many opportunities to play professional basketball in the United States compared to women, a total of 450 roster spots across the thirty NBA teams compared to 144 roster spots on the twelve WNBA teams. In Major League Baseball, each of the thirty teams carries 26 athletes on their regular-season rosters, providing opportunities for 780 players, not including the minor leagues. In softball, before the National Pro Fastpitch folded in 2021, there were approximately 100 players. A new professional softball league, Women's Professional Fastpitch, was announced in 2021 and played an exhibition season in 2022 with two teams.

It can be tricky to recruit enough women athletes to make up a meaningful cohort of participants, especially for studies that compare women and men. Louise Burke is the chair of sports nutrition at Australian Catholic University and a leading sports nutrition researcher. At a recent academic conference on women in sports, she laid out the challenges: "What do you match them for? Their lean mass? Their VO_2 max [a tool commonly used to measure cardiovascular fitness]? How much training they do? It's very difficult to find the features or characteristics that you can say, 'I'm pretty sure these people are the same except for their sex and therefore I can compare [sex differences].'"

More critically, funding drives everything. Outside of government entities like the National Institutes of Health, the key supporters of exercise science research are sports governing bodies like the NCAA, the National Football League, and the Federation Internationale de Football Association (FIFA) along with private companies like Gatorade. On the whole, these institutions prioritize men's sports over women's sports, which is reflected in their allocation of resources. In 2022, USA Today analyzed how 107 schools in the highest level of Division I sports spent money on travel, equipment, and recruiting in sports with comparable women's and men's teams during the 2018–2019 and 2019–2020 seasons. What they found was that for every dollar spent on the men's teams, schools spent 71 cents on the women's teams, adding up to an additional $125 million more for men's sports over two years. FIFA raised the prize money for the 2019 Women's World Cup to a total of $30 million, with $4 million going to the winning team, but that's still less than 10 percent of what they allocated to the men's tournament. It's no surprise that when it comes to earmarking resources for research purposes, these institutions direct their attention primarily to men. With revenues from women's sports estimated to be a tiny sliver of the $481 billion global sports market, there often isn't money to support research activities at the same level.

With a funding pipeline and prevailing sociocultural norms set up to support and advance men, a paradigm developed: scientific research became centered on the "male specimen." The idea of the male norm solidified further with the creation of the "standard man." Following World War II and the proliferation of nuclear technology, the International Commission on Radiological Protection (ICRP) wanted to know how much radiation a worker could be exposed to. In September 1949, the group met in Ontario, Canada, and created the parameters for an average worker, who at the time was typically a man, based on tissue samples and organ weight gathered from autopsies and the normal daily patterns of someone living in a temperate climate. They outlined everything from his body weight (70 kilograms) to his daily water intake (2.5 liters from liquid and food) to his body's overall water content (50 liters). In 1974, the "standard man" was revamped to reflect the general population and renamed the "reference man." He was described as a Caucasian man between twenty and thirty years old, five foot six inches tall, who lives in Western Europe or North America.

The ICRP didn't intend for the reference man to represent the average person; he was meant to be modified depending on the characteristics of different populations. Scientists needed a benchmark with precisely defined traits so that if they were to recalibrate the recommended radiation exposure levels, they had a basis for their calculations. But the idea of individualized adjustments was largely overlooked and forgotten.

Other domains like biomedical research and product design began to adopt the reference man. His influence can be felt everywhere from the cold temperatures in offices to the design of personal protective equipment and strength training guidelines. To scientists, engineers, and designers, any differences between men and women, outside of reproductive organs and functions, could be explained as just differences in scale—smaller organs, smaller stature, and different body composition and weight. In other words, women are just smaller versions of men.

This view isn't surprising when you consider that men primarily conduct research, dating back to the days when figures like A. V. Hill and David Bruce Dill laid the groundwork for exercise physiology research. Even today, female physiology isn't always taught in university courses. A case study of sports-related courses from one UK university found that men athletes and male bodies were normalized across the curriculum, ranging from subjects like sports management to applied sports science to physical education. Gender and women's physiology were considered only superficially, even in anatomy and physiology modules. They were more likely to be discussed in sociology and social science modules. Not only does this influence the skills of future researchers, professionals, and practitioners, it can also impact how students think about and frame the role and place of women in sports and science.

I have to admit, when reading scientific papers, I used to skip over the methodology section. I wanted to get to the good stuff, the results and the discussion. As a non-scientist, I didn't necessarily care about the details of how a study was set up. I wasn't going to replicate it in a lab and I generally assumed that researchers used the most up-to-date and rigorous procedures. If they didn't, why would the study be published?

But as I became immersed in the literature, I realized that methodology matters—a lot. It creates the universe in which we understand research findings and what they potentially mean across broader populations. It also creates the foundation that can perpetuate a specific point of view. Without examining how studies are established and the assumptions underlying them, we can't identify where the bias comes from. If we can't identify the source of bias, we can't change the system.

In research, scientists tend to favor simplicity, especially when the goal is to construct models to explain complex physiological, molecular, and chemical behavior. They strive to eliminate as many extraneous factors as feasible to reduce any "noise" in the data. Ken O'Halloran, professor of physiology at the University College Cork in Ireland, told me that scientists choose to use homogenous strains of male rats of the same age to minimize variability in their studies. "Therefore, when you're looking at some intervention, fewer observations are required to detect a real difference," he says. The same reasoning applies to studies involving humans: Scientists look to recruit a standardized cohort of participants, one that's relatively controlled and with few external variables that may influence the outcome of the study.

Women's bodies stand in direct opposition to the need for simplicity. When I asked experts why more women aren't included in research studies, I kept hearing the same answer: women are complicated. The biggest consideration is the menstrual cycle, the ebb and flow of female sex hormones, like estrogen and progesterone, that regulate the reproductive system. But these chemical messengers also influence a wide range of other physiological functions. A person's hormonal status and the health of their menstrual cycle—such as if they have a regular, irregular, or absent cycle and whether they take hormonal contraceptives—can affect factors such as metabolism, muscle function, hydration, temperature regulation, central nervous system fatigue, and respiration. These hormonal rhythms also change throughout a woman's life.

It's understandable then that when researchers want to understand the specific way the body functions or how the body adapts to a training or nutrition intervention, an ever-changing hormonal environment can throw a wrench into a scientist's best laid plans. The menstrual cycle— as well as hormonal contraceptive use and menopause hormone therapy—adds additional elements scientists need to control for in their research design. Say you're studying heat adaptation in women athletes and find that subjects experience a 0.5- to 1-degree drop in body temperature over the course of the study. You may conclude that the athletes acclimated to the heat, but it depends on where they were in their cycle at the start and end of the study. Body temperature changes throughout the month; it's generally lower during the first part of the cycle, rises with ovulation, and remains elevated until it drops again at the next period. If you happen to take a baseline body temperature around ovulation and again after the participant has started her period, the drop in body temperature could be the body's normal rhythms rather than a demonstration of heat adaptation.

It takes a series of steps and some methodological pre-planning to properly account for the menstrual cycle's hormonal rhythms, establish a solid study design, and evaluate and interpret the results. When possible, exercise physiology and nutrition professor Anthony Hackney prefers to pre-recruit subjects, ideally six months before the start of a research project. He asks participants to track their body temperature for a few months, which gives him an overall picture of a person's hormones and whether their cycle is consistent and "normal." If the cycle appears regular, he then sends them home with urinary ovulation kits. Like a pregnancy test, the ovulation kit measures the levels of certain hormones in urine. Once a day, starting a few days before the midpoint of their cycle, participants pee on one of the dipsticks to confirm ovulation. "It tells me that what I saw with the temperature changes is actually occurring within the endocrine system inside the body," Hackney told me. He also checks hormone status during the study via blood or urine samples.

The problem is that not every scientist wants to go through the hassle of pre-planning or taking extra steps to verify hormone and menstrual cycle status during the course of a research project. For one, it can make it difficult to plan and schedule the study. If scientists want to focus on a specific phase of the cycle, that means women can only participate during certain days of the month—which may or may not be convenient for the researchers or participants, particularly if they're athletes. "It's hard enough getting elite athletes to be part of these things in the first place, but if you're saying I can only have them in this part of the menstrual phase to make that standardized [across the study], how am I going to organize that? It'd be easier to say, 'Move the building an inch to the left,'" Louise Burke says. Tests to monitor hormone levels also mean additional expenses, on top of all the normal costs of equipment, staff, and recruitment. Hormone levels in men are mostly constant, making it more straightforward to study male biology in cells, animals, and people. Scientists often choose to sidestep the extra logistical hurdles and default to recruiting only men so they can get on with their work.

The fault, however, doesn't always land on the researcher. Volunteer bias—who chooses to participate in research studies and whether the pool of volunteers is representative of the target population—can play a role too. When we talk about underrepresentation of women in sports science research, we assume that women and men are equally invested and willing to volunteer their time and bodies. While women and men haven't been formally surveyed about why they do or don't participate in these types of studies, there are other factors that influence who signs up and shows up at the lab, including interest in the topic at hand, schedule availability, and pain and risk tolerance. For example, in her over forty years of research, anecdotally, Burke has found that women are more reluctant to participate in nutrition studies where they have to change the way they eat. With men, on the other hand, she says, "it's just calories down the gob for them," and they sign up with minimal concerns.

There's no denying that the procedures needed to account for and control for the menstrual cycle, hormonal contraceptive use, and menopause hormone therapy in scientific research involve extra work. But the mere fact that a person menstruates has become a convenient escape hatch. It's a blanket excuse that scientists can use to duck out of studying women—even when hormones don't play a role in the biological processes under consideration. Ultimately, that leaves topics unrelated to cisgender men understudied.

When Ken O'Halloran reviewed original research published between January 2017 and December 2019 in *The Journal of Physiology*, he found that studies involving healthy young men dominated the literature. There were three times as many single-sex studies involving men compared to single-sex studies involving women. Women-only investigations predominantly probed issues like pregnancy, menopause, and reproductive disease. Even though there's been an increase in mixed-sex studies in recent years, male bias persists, with many studies where men participants outnumber women four to one.

The data gap becomes even more skewed when it comes to exercise and sports science. In studies published in the three top sports science and exercise medicine journals—*Medicine & Science in Sports & Exercise, British Journal of Sports Medicine,* and *The American Journal of Sports Medicine*—between 2011 and 2013, only 39 percent of the total participants were women and 9 percent of studies were carried out with only women. Seven years later, not much had changed. A team of researchers from the United States and United Kingdom reviewed more than five thousand studies published in six leading academic journals between 2014 and 2020, involving twelve million participants. The overall number of participants who were women edged down to 34 percent, and just 6 percent of studies were conducted exclusively with women. What's more, of the research that included only men, just 0.6 percent of those studies investigated a topic unique to men. That means women could have been included in the remaining 99.4 percent of studies but weren't.

It's not surprising, especially when you consider that the majority of those publishing in the sports science space are men. Between 2000 and 2020, women accounted for only one-quarter of first authors and roughly one-sixth of senior authors of randomized control trials—the gold standard of scientific studies—in top sports science journals. (The first author is a coveted position and denotes who contributed the bulk of the work of the study. The senior author is typically the head of the lab, research group, or department.) Implicit gatekeeping may also keep research on women out of journals: At top-ranked sports science and rehabilitation journals, women occupied less than 25 percent of editor positions and only 10.4 percent of editor in chief positions—those responsible for prioritizing a journal's research topics. Editors from low- and middle-income countries, particularly in South Asia and Africa, were underrepresented too.

Even some of the most prominent researchers in the field don't realize how big the data gap is. In 2018, Louise Burke conducted an audit of her own studies. Of sixty-eight studies that investigated performance-related sports nutrition interventions, ten included female and male participants, two compared outcomes between women and men, and only three studies exclusively targeted women. "I was embarrassed. It's me that's contributing to the underrepresentation of females in sports nutrition research," Burke says.

Yet, in her work with Australian Olympic athletes, Burke is judged by metrics like medal counts, and women are among the biggest contributors to her country's medal haul. Australian women are 35 percent more likely to win gold compared to the men. "It's a bit silly that we've been spending so much time on the less successful people," Burke says in reference to men athletes. "That's when I started thinking this is a really major area that I haven't done a good job in. If I can put my hand up and say I need to be better, maybe that's part of the advocacy that's needed to get change."

The exclusion of women from scientific research has real implications. By making research appear easier to do on the surface, we eliminate an important biological variable from cell, animal, and human studies. When we base our ideas only on what we see and find in men, it can create a sampling bias that distorts our understanding of what's considered "normal" physiology across whole populations. As a result, we don't have a firm grasp on the scope of non-male-specific phenomena, the underlying mechanisms at play, or the true magnitude of differences—and similarities—between women and men. As research on women and men becomes more equal, we will start to get a truer sense of what the findings mean and how they can be applied across entire populations.

Among the trailblazers to study the effects of exercise on women was Barbara Drinkwater, the first woman elected as president of the American College of Sports Medicine in 1988. Born in 1926 in Plainfield, New Jersey, Drinkwater loved sports but grew up at a time when there weren't many opportunities for women to go on to higher levels of competition. Instead, she pursued an undergraduate degree in physical education at New Jersey College for Women (now known as Douglass College and part of Rutgers University) and went on to work as a women's high school basketball coach, swim instructor, and physical educator.

Drinkwater was a feisty advocate for women. She was intent on showing that women are up to any task, especially when it came to sports, and wasn't afraid to stand up for her beliefs. After completing her master's degree, she took a teaching position at the University of Nebraska. Always the athlete, she played rec volleyball and basketball through the local YWCA, but her department chair called her aside and told her that members of the department didn't play sports, a rule that applied only to women staff members. Drinkwater left at the end of the year to pursue her PhD at Purdue University.

Drinkwater was unique in the male-dominated world of exercise physiology. She didn't know much about the field before arriving at the Institute of Environmental Stress at the University of California, Santa Barbara, in the 1960s. The research university and publish-or-perish ethos was different from the world of physical education Drinkwater was accustomed to. But she was good at statistics and was responsible for data analysis and experimental design for the institute's studies. Over time, she learned on the job under the tutelage of the best in the field, including institute director Steven Horvath, who previously worked in the Harvard Fatigue Lab under David Bruce Dill. Drinkwater soon began conducting pioneering research of her own on women.

Drinkwater's work coincided with the groundswell of women engaging in physical activity and competitive sports in the 1960s and 1970s. As more girls and women began to play sports, doctors noticed something peculiar. While there were numerous positive health benefits of exercise, physicians also observed altered or delayed menstrual cycles in high school, college, and Olympic athletes compared to nonathletes, a condition then called "athletic amenorrhea." These athletes also experienced a fair number of injuries, particularly stress fractures, which can be a sign of poor bone, menstrual, and nutritional health and can leave athletes at risk for additional fractures in the future. It was a crucial discovery that sent scientists scrambling to understand how and why exercise disrupts reproductive function and bone health. Drinkwater published two seminal studies in well-respected medical journals—in 1984 in *The New England Journal of Medicine* and in 1986 in *The Journal of the American Medical Association*. These papers sent a message: women face unique challenges when it comes to sports and deserve attention.

Aurelia Nattiv, a professor of sports medicine, family medicine, and orthopedics and team physician for UCLA Athletics, remembers reading Drinkwater's papers in her journal clubs in medical school and during residency in the late 1980s. Drinkwater's science was meticulous, and Nattiv told me she studied those papers "like they were the Bible." Nattiv's interest in exercise, nutrition, and women's health led her to a primary care sports medicine fellowship at UCLA, one of the few women in the country to pursue the specialty at the time. When she arrived at UCLA, women athletes flocked to her. They hadn't worked with a woman sports physician before and told Nattiv about their diets, menstrual cycles, and other issues they weren't necessarily comfortable sharing with doctors who were men.

Pretty soon, Nattiv noticed a pattern. The athletes who had low bone density and were getting stress fractures also weren't eating enough food to keep up with the demands of their sport. Or they exhibited patterns of disordered eating. They tended to have problems with their menstrual cycle too. At an American College of Sports Medicine meeting in the early 1990s, Nattiv shared her findings with a group of colleagues and found similar patterns among women athletes cropping up in other parts of the country.

Among those at the meeting was Drinkwater. The reports from Nattiv and other clinicians mirrored what Drinkwater found in her studies. Nattiv says, "Dr. Drinkwater basically gathered us together and took us by the hand and said, 'You guys have something here.'" In a brainstorming session, Drinkwater and Nattiv, along with colleagues Rosemary Agostini and Kimberly Yeager, coined the term "female athlete triad" for the condition that, at the time, included disordered eating; amenorrhea, or the absence of menstruation; and osteoporosis.

The identification of the female athlete triad was a pivotal moment for the field of women's health. With Drinkwater's encouragement and mentorship, Nattiv and Agostini organized a meeting in Washington, DC. Over the course of two days, scientists, clinicians, physicians, Olympic athletes, and members of Congress gathered to discuss the science behind the female athlete triad and the clinical presentation of the condition. After the meeting, Nattiv was whisked away in a limousine to a local CNN affiliate, where she was interviewed by Connie Chung, bringing a national spotlight to the triad for the first time. In 1993, Nattiv, Drinkwater, Agostini, and Yeager published the first paper on the triad. The increased attention spurred a call to action for more research, along with the need for guidelines to help clinicians prevent, identify, and treat the triad in women athletes.

Interest in the triad intensified during the summer of 1994 when world-class gymnast Christy Henrich died from complications due to anorexia, eight days after her twenty-second birthday. Henrich's nickname was E.T., for "extra tough," and she pushed her body to the limits in an effort to be the best gymnast in the world, including underfueling her body. Drinkwater and Nattiv, who was involved with USA Gymnastics at the time, were both crushed by the news. In a 2017 interview, Drinkwater said, "Recognizing the reality of the triad was shocking. I would say I was angry, perhaps even 'mad as hell,' when I [would] see coaches requiring female athletes to do things that might have lifelong consequences, but they worried more about winning than the well-being of the athlete. We had kids dying. There's just no excusing that."

Investigation into the female athlete triad became all-consuming for many sports medicine physicians and sports scientists. "It overshadowed any other female physiology and exercise-related work for a long time," says Anthony Hackney. As other researchers became interested in studying female physiology and entered the field, they pursued this research pathway. This may in part explain why it's taken a while for exercise physiologists and others to look beyond how exercise impacts reproductive function and bone health and to ask a broader question: How does the unique physiology of the female body, including its kaleidoscope of hormones, affect exercise capacity and athletic performance? There was also an underlying concern among some women's sports activists and medical professionals that highlighting sex differences in sports and exercise might shine a bad light on women athletes. If physical activity was shown to have a downside for women, women could lose their hard-won privilege of playing sports and shy away from exercise altogether.

The scientific community is moving toward greater representation in biomedical research, thanks to shifting infrastructure and public policies. The seeds were planted in 1985 when the Public Health Service Task Force on Women's Health first acknowledged the gap in data regarding women's health and recommended that "[clinical] research should emphasize disease unique to women or more prevalent in women." In 1993, the National Institutes of Health (NIH) Revitalization Act mandated the inclusion of women and minorities in NIH-funded clinical research, one of the nation's biggest sources of research funding.

Still, it's been a slow road. It wasn't until 2001 that the Institute of Medicine (now called the National Academy of Medicine) concluded that differences between the sexes do have important consequences when it comes to scientific investigations. More recently, NIH issued a policy requiring the inclusion of sex as a biological variable in preclinical research in vertebrate and human studies. Since 2016, scientists have been required to account for sex—just like age and other variables—and explain how it is considered in a study's design, analysis, and reporting. If scientists propose a single-sex study, they must explain their rationale, drawing from scientific literature and other evidence. The guidance states, "Absence of evidence regarding sex differences in an area of research does not constitute strong justification to study only one sex."

Research journals are paying attention now, as well. Publication is the main way scientists share their scholarly work, not to mention that it also provides critical notches on a career ladder. More and more, journals recognize that the exclusion of sex in a study's analysis is a limitation and some expect authors to address these issues in the paper. Ken O'Halloran, who served as one of *The Journal of Physiology*'s senior editors from 2019 to 2022, told me he wants the research featured in its pages to apply across the spectrum of the human population. Researchers can't just cherry-pick a subset.

Though policy and systemic changes have started to take root, it will take time for scientific output and the research base to catch up. There is research on female physiology and biology, but it lacks the nuance and depth of the evidence we currently have on men. "Every single day, I'm trying to help guide decision-making for female athletes and their coaches based on very shaky, limited data," says Trent Stellingwerff, who works with elite Canadian athletes and leads innovation and research for the Canadian Sport Institute Pacific. "There's a big gap and there are big assumptions made at times."

While some theories may hold up across women and men, Stelling-werff told me there may be situations when a subtle shift in the application of a training or nutrition strategy could be incredibly important for how an athlete feels, trains, and recovers. At the elite level, where races and matches are won by the slimmest margins, it could be a very big deal. By not thoroughly exploring the data gap between women and men, we miss opportunities to optimize performance and improve the health and longevity of millions of girls and women in sports around the world.

Take the example of beetroot juice—in the sports world, the jewel-red liquid is touted for its endurance-boosting powers because it's rich in nitrates. The body converts dietary nitrates from beet juice and other sources like green leafy vegetables into nitrite and, ultimately, nitric oxide. Nitric oxide is thought to influence important functions like blood pressure, muscle contractility, and oxygen utilization in ways that may enhance performance. But the enthusiastic endorsement of dietary nitrate supplementation is based primarily on evidence found in men. In one review of existing research, more than one hundred studies were conducted exclusively with men as participants. Only seven studies were conducted strictly with women.

It turns out, women may metabolize nitrate differently than men. Despite no real difference in the bacteria or other characteristics inside the mouth, women are better at converting nitrate to nitrite, the first step in the pathway that leads to nitric oxide. While this may mean women are more responsive to nitrate supplementation, it could also mean that women are already optimized to reap the most out of dietary nitrate and that, in fact, men would benefit more from beetroot juice. At the moment, there isn't enough information to determine who would best benefit from supplementation. It's only by doing more studies that scientists will begin to unravel these types of mysteries.

———————

Scientists can make their research more inclusive and representative if they are willing to commit to it. In recent years, several leading experts in the sports science community have led a charge to propose common language around research design methodologies and protocols. Their goal is not only to help researchers feel less intimidated when it comes to studying women, but to improve the quality of the research too. After all, the field won't progress without good science, and there need to be more studies that are designed for women, that use women as participants, and that are interpreted and applied through a women-specific lens.

And paradigms are starting to shift, especially as more women— particularly those who played sports in their youth—move into careers in scientific and medical fields. "We're finally seeing more women scientists have the opportunity to do the work and pursue those things they are passionate about," says Anthony Hackney. It will take continued hard work, but we're headed toward a watershed moment.

Kate Ackerman is one of the women whose interest in sports, physiology, and medicine converged around issues related to women athletes. She didn't necessarily expect to become an athlete herself. She told me she was more of a musical theater person in high school, but when she was turned down by two a cappella groups at Cornell University her freshman year, she decided to try something different. She headed down to the boathouse and walked on to the crew team. She eventually worked her way up to rowing varsity for Big Red and made the U.S. national team while in medical school. With her burgeoning understanding of human physiology, Ackerman wondered if she and her teammates should train any differently from men. She pored through the scientific literature and came up empty. "I realized there wasn't any specialized approach. There was just a catchall, and it was based mostly on men," she says.

Ackerman knew then that this was an area that deserved more attention and she wanted to carve out a career that married athletics and medicine. While most sports medicine practices tend to treat patients in a silo and focus primarily on orthopedics and musculoskeletal injuries, Ackerman approached the problem from a different angle. In 2013, she established the Female Athlete Program at Boston Children's Hospital, which takes an interdisciplinary approach to treating athletes. Shortly after, she started the Female Athlete Conference, a biannual conference that brings together researchers and practitioners from around the world to learn and collaborate. Conference proceeds are used to fund additional investigations to address the funding gap for women-specific research. "We're here. We're doing sports. It's better that we understand the issues so we can help put women in the position to be successful rather than being afraid to find out the differences. Let's find them out and let's address them," Ackerman says.

Athletes themselves are demanding better information and helping to spread the word about why it matters. Pro mountain biker Kate Courtney told me, "It really changes how you interpret data if you're using data that is mostly based on male athletes." It can make decision-making more challenging "because of the lack of data on women specifically." It's a data gap Courtney thinks is underappreciated, even by women athletes.

While the exclusion of women from scientific research appears to be a case of omission, at least on the surface, gender bias and sexism play a more overt role in another realm. Myths about women's bodies have long been used against women to hold them back from sports. To understand the impact of these widespread misconceptions, we need to understand where they came from and interrogate what we do—and don't—know about sex and gender.

2

More Than Just Hormonal

• • • ● ● ● • •

T he idea that women are the weaker sex has long been deeply ingrained in Western cultures. Women were seen as the foil to men: the delicate, feebleminded, and feminine counterparts to male virility, wisdom, and masculinity.

Ever since the time of Greco-Roman antiquity, the source of a woman's weakness centered squarely on her physiology and anatomy. It was assumed that features like a womb, ovaries, and breasts somehow made women inherently less capable of pursuing the same activities as men. Doctors in ancient Greece even believed that the uterus wandered the body like an animal, causing all sorts of maladies along the way.

Ironically, while a woman's reproductive system was seen as the source of her vulnerability, her capacity to bear children was central to her role and value in ancient Greece and Rome. It was imperative to protect her ability to procreate. Physiology became a tool for policing women's bodies and barring them from different realms of public life. All activities, especially physical activity and sports, were evaluated based on their potential to harm a woman's reproductive function.

These concerns laid deep roots for the idea that physical activity was incompatible with the female body.

It's no wonder then that women who choose to be physically active and participate in sports have always been viewed with a healthy dose of skepticism. The very idea of women competing in sports was abhorrent to the ancient Greeks and Romans. Sports was an arena to celebrate men and the male form. In *The Republic,* Greek philosopher Plato mused that if women and men were treated equally, "the most ridiculous thing of all will be the sight of women naked in the palaestra, exercising with the men, especially when they are no longer young; they certainly will not be a vision of beauty." Roman satirist Juvenal similarly expressed his distaste for the idea of women exercising or training as gladiators, writing, "What modesty can you expect in a woman who wears a helmet, abjures her own sex, and delights in feats of strength?"

While men tested their athletic prowess across a range of sports in the ancient Olympic Games, from running to discus to wrestling, women were barred from competing. Instead, Grecian women hosted their own competition. Known as the Heraea or Heraean Games, it was held every four years in honor of the goddess Hera and limited to a series of footraces. Even then they were given a handicap. The men's race covered the full distance of the track, but the women's course was shortened by one-sixth in length. (Sparta, however, was the exception to the rule in ancient Greece. There, women were encouraged to take up sports because Spartans believed athletic women gave birth to stronger children who would better defend the city-state in the future.)

During the Renaissance, the separation between women's and men's physical training was further delineated in the *Libro del Ejercicio Corporal* (or *Book of Bodily Exercises*). Written by Spanish physician Cristóbal Méndez and published in 1553, it was the first book devoted wholly to the subject of exercise. Méndez dedicated an entire chapter to women's exercise, drawing a clear distinction between women's physical activity needs based on social class. He claimed that rural and working-class women required only the activities of daily life—plowing, mowing, cleaning, and caring for children. Upper-class women of leisure, on the other hand, were advised to walk around their home or estates and yell at their servants for exercise.

While Méndez's recommendation of yelling as a form of exercise seems absurd, it was actually based on his observations during a trip from Europe to the court in the territory of New Spain (where Mexico City was its capital). He encountered nuns, who spent the entirety of their days praying and singing, and he assumed that those activities were the reason for their good health. However, as University of Texas professor and sports historian Jan Todd told me, "He misses the fact that they're not getting pregnant and dealing with all of the medical problems related to that." This was the real reason they lived a long, healthy life, not the vocal exercises. Yet this idea left an enduring impact on women's physical culture for years to come. In the 1600s, English writers Francis Beaumont and John Fletcher claimed that "as men do walk a mile, women should talk an hour after supper . . . 'tis their exercise." In the 1700s, Nicolas Andry, a doctor, believed that since women "are more loquacious," they didn't need as much exercise as men.

In the nineteenth century, the medical community continued to cinch in the boundaries of what women could and could not do. This time, the theories were based on the idea of vitalism, that human beings have a fixed, finite amount of energy and any nonessential activity, like exercise, shifted energy away from essential systems. Physicians believed women used up part of their vital energy every month during menstruation, leaving them with an energy deficit and periodic weakness that men weren't subject to. In his 1873 book *Sex in Education; or, A Fair Chance for the Girls,* Boston physician Edward Clarke wrote that any activity undertaken during women's periods would leave them prone to sickness—or worse, it could deprive their reproductive organs of nourishment and leave them unable to bear children. Instead, women should rest to protect their delicate constitutions.

Doctors have also long believed that vigorous movement could loosen the uterus and cause it to fall out of the body. In 1892, Senda Berenson Abbott, a physical educator at Smith College, adapted the game of basketball for women. Players were to stay within one of three zones on the court and were only allowed to dribble the ball three times. These rules were designed to force players to pass the ball and play more collaboratively to avoid excessive exertion and competitive impulses that were considered "unladylike." Women were also discouraged from exercising because doctors thought it would inhibit the development of the pelvis and lead to taut abdominal and pelvic muscles, making childbirth more difficult. Breasts, too, were seen as another reason why women weren't cut out for sports. In an article in *The New York Times* at the end of the nineteenth century, the reporter wrote that women were bad swimmers "because they have breasts where men have pectoral muscles."

At the turn of the twentieth century, Pierre de Coubertin dreamed of a modern Olympics, a competition he described as "the solemn and periodic exaltation of male athleticism, based on internationalism, by means of fairness, in an artistic setting, with the applause of women as a reward." Coubertin didn't believe women were suited for sports, saying, "No matter how toughened a sportswoman may be, her organism is not cut out to sustain certain shocks." While women were allowed to participate in the second Olympics, their presence was minuscule. Twenty-two women competed in five sports—tennis, sailing, croquet, equestrian, and golf. In contrast, 975 men competed in twenty sports. Years later, Coubertin still believed that women had no place in the Olympics, except to crown the victors.

Pseudoscientific assumptions about women's bodies persisted well into the twentieth century. When women made their debut in track and field at the 1928 Summer Olympics in Amsterdam, they competed in five events: the 100 meters, the 4×100-meter relay, the high jump, discus, and the 800 meters. The most controversial was the 800 meters. Officials and medical staff were concerned that running twice around the track would be too much for the women to handle. But when the athletes lined up in the Olympic Stadium, the stands were packed and the top three women—Lina Radke from Germany, Kinue Hitomi from Japan, and Inga Gentzel from Sweden—all broke the previous women's world record. After crossing the finish line, some competitors lay down on the infield grass to catch their breath.

The media, however, had a different perspective on the record-setting achievement, claiming that the event was alarming to watch. In the *Chicago Tribune*, the headline read, "5 Women Track Stars Collapse in Olympic Race," and reporter William Shirer wrote that when American Florence MacDonald crossed the finish line in sixth place, she "collapsed, falling onto the grass unconscious. Officials had to work over the Boston school girl for several minutes before she could regain strength enough to stand up." Another renowned sportswriter claimed that five of the eleven "wretched women" dropped out of the race while five others collapsed once they crossed the finish line.

Newspapers across the globe spun the same false narrative. Nine athletes, not eleven, raced that day, and no one collapsed at the end of the race. Still, coaches, trainers, the International Amateur Athletic Federation (IAAF), and International Olympic Committee (IOC) officials latched on to this storyline as proof that women were ill-suited for strenuous competition. Officials nixed the women's 800 meters from the Olympic program, once again using women's supposed frailty as an excuse. For the next thirty-two years, women were barred from competing in races longer than 200 meters. The 800 meters wasn't reinstated at the Olympics until the 1960 Summer Games in Rome.

When swimmer Greta Andersen blacked out during the preliminary heats of the 400-meter freestyle at the 1948 Summer Olympic Games in London, her body was blamed. The twenty-one-year-old Danish swimmer held the best time in the world in the event and had won gold a few days earlier in the 100-meter freestyle. During her heat, Andersen led the race until, suddenly, her legs stopped working. She felt paralyzed and sank to the bottom of the pool. Amid the commotion, a Hungarian swimmer on the sidelines noticed Andersen in distress. He dove into the water and rescued her.

It turned out that Andersen was supposed to start her period the day of the preliminary heats. The team doctor worried that it would affect her chances of winning, so he gave her an injection to delay her period. Trusting that the doctor knew best, Andersen didn't protest. While she never found out what was in the injection (even many years later), Andersen believed it caused her to black out. But to her coaches, Andersen's fainting episode confirmed their suspicions that a woman's period is detrimental to athletic competition.

Even as doctors and scientists learned more about the inner workings of the human body, many still believed women were inherently inferior to men simply because their bodies were different. In the book *Physiology of Strength*, published in 1961, physician Theodor Hettinger put forth the notion that women were generally only two-thirds as strong as men and only 50 percent as responsive to training. In her memoir *Marathon Woman*, Kathrine Switzer, who in 1967 became the first woman to run the Boston Marathon as a registered athlete, remembers interviewing the girls' high school basketball coach for her school newspaper and asking if girls would ever play the same version of the game as the boys. She recalled the coach saying no because the large number of jump balls "could displace the uterus." At the first FIFA Women's World Cup in 1991, when the women took the pitch, they only played eighty minutes instead of the full ninety-minute game.

Years later, U.S. captain April Heinrichs explained skeptically, "They were afraid our ovaries were going to fall out if we played ninety."

Sadly, at the turn of the twenty-first century, the overriding beliefs about women's bodies and athletic capability hadn't changed. Women were still barred from ski jumping as recently as the 2010 Olympics. Gian Franco Kasper, former president of the International Ski Federation, justified the decision in a 2005 interview, saying that ski jumping "seems not to be appropriate for ladies from a medical point of view." The prevailing concern was that landing from a ski jump might cause the uterus to burst. Women were finally granted their own ski jump event in 2014 at the Winter Olympics in Sochi, Russia. Women's physical inferiority was even invoked in the U.S. Women's National Soccer Team's equal pay lawsuit in 2020. In its legal filing, U.S. Soccer Federation claimed that scientific evidence proved that players on the women's team were physiologically inferior to those on the men's team. Men were inherently faster and stronger, and the men's game required more skill than the women's game, thereby justifying the pay discrepancy.

At the four major Grand Slam tennis tournaments—the Australian Open, Wimbledon, the French Open, and the U.S. Open—women play a best-of-three-sets match format while the men play best-of-five. Maintaining a two-tier tennis system at the highest level of the sport perpetuates the idea that women aren't as capable as men. Women athletes have expressed their willingness to play the longer-format matches as far back as 1901, but tennis officials haven't taken them up on it except for short-lived exhibition games.

Given the dogged perception that women are the lesser sex, it's no wonder women tend to pretend that their anatomy and physiology don't come into play in the sporting arena. While women's bodies and their ability to bear children have been used to constrain their participation in physical activity, there is no real evidence to support claims that exercise adversely affects a woman's reproductive capacity and childbirth outcomes. Unsurprisingly, it was women who sought to dispel the myths that "biology is destiny" and women are the "weaker sex." They battled those ideas the best way they knew how—with science.

In the late nineteenth century, Mary Putnam Jacobi was a living testament that education and physical activity weren't detrimental to a woman's health, as Edward Clarke had claimed in *Sex in Education*. She was a highly regarded physician and one of the few women practicing medicine in the late 1800s. She earned a degree from the New York College of Pharmacy, a medical degree from the Female Medical College of Pennsylvania, and a second medical degree from the École de Médecine at the Sorbonne in Paris (as one of the first women to attend the institution). She was also the mother of three children.

Jacobi was a stickler for rigorous scientific research, which is why she took issue with Clarke's claims. To Jacobi, Clarke's case seemed to rest solely on the belief, rather than any actual evidence, that being identified as a woman at birth was the root cause of all problems. In her manuscript *The Question of Rest for Women During Menstruation*, written in response to Clarke's book, she argued that the beliefs put forth by Clarke and others about women's inherent frailty "should only be accepted after the strictest scrutiny."

So Jacobi proceeded to scrutinize. She researched women's experience of menstruation along with their health, education, exercise habits, and any debilitating symptoms they experienced. She conducted a thorough review of the research literature, surveyed 268 women, and ran experiments with a smaller group of women. She measured physical indicators like pulse and temperature, collected urine samples, and tested muscle strength. She concluded that "there is nothing in the nature of menstruation to imply the necessity, or even the desirability, of rest." Menstruation didn't deplete a person's health or energy. Rather, she found that women were healthier when engaged in activity. Her work earned her the Boylston Prize at Harvard University in 1876, making her the first woman to win the prestigious award.

In Germany, another group of women doctors challenged long-standing paternalistic assumptions about women's bodies by setting forth scientific fact over subjective opinions. In the early twentieth century, Alice Profé, a physician in Berlin and high-ranking member of the Central Commission on National and Youth Games, argued against dividing physical education by sex. She stressed that the physical bodies of women and men, from blood to muscles to the respiratory system, were more similar than different. She explained that the perceived weakness in women wasn't due to their reproductive biology or physiology alone—it was due to a multitude of external factors, such as restrictive clothing, lack of education, and insufficient physical activity. For example, a study from 1924 showed that while shallow chest breathing is typically associated with women, it wasn't a trait *inherent* to women. Instead, it was the result of social customs like corset wearing. Unsurprisingly, when your ribs are constrained and unable to expand, your breathing is very likely to be confined to the upper part of the chest.

Soon, other women doctors in Germany began to investigate if and how sports and physical activity impacted the body, menstruation, and childbirth outcomes. By 1934, more than 120 scientific studies—involving approximately ten thousand active girls and women—were published in Germany alone, and no study found any scientific reason to restrict women's participation in active pursuits. There was no evidence that exercise harmed women or affected them differently than men. Yet, despite the scientific rigor of the studies, the male-dominated medical community rejected these claims. The reason? Women couldn't speak objectively about women's issues or their own sex.

The legacy of Jacobi and Profé resurfaced in the second half of the twentieth century, this time in the field of kinesiology. When Barbara Drinkwater began her research career as an exercise physiologist at the Institute of Environmental Stress at the University of California, Santa Barbara, in the 1960s, she developed hypotheses about girls' and women's health and physical performance that challenged existing data and historical assumptions. Drinkwater came across exercise physiology textbooks from the 1950s that stated a woman's capacity for aerobic exercise declines beginning at age fifteen, effectively dissuading girls from pursuing physical activity once they hit adolescence. When Drinkwater dug into the research studies underlying these claims, she noticed that the scientists had enlisted inactive women. "These women had been discouraged from ever being active after age fifteen, so naturally there was no indication that they could do more," Drinkwater reflected years later in an American College of Sports Medicine interview. If they weren't encouraged to be active, how could they be expected to be physically fit or strong?

Drinkwater brought an exactingness to examining many of the pseudoscientific beliefs that had long held women back from physical activity. For example, one argument against women running long distances was that they couldn't cope with the heat stress associated with the sport. Again, Drinkwater went to the library to examine the studies. She found that researchers had recruited sedentary women and compared them to young, healthy men—people who were likely to participate in sports, giving them a leg up in cardiovascular fitness. The two groups were given the same absolute workload despite their differences in aerobic capacity and fitness, and, unsurprisingly, the men performed better compared to the women.

Drinkwater decided to run her own studies. With the running boom taking hold in the 1970s, she had access to a growing pool of trained women marathon runners. When she compared the runners against a control group of women, she found that the ability to regulate and tolerate heat had nothing to do with whether athletes were identified as female or male at birth, and everything to do with cardiovascular fitness. The women runners exhibited lower heart rate, skin temperature, and internal temperature, signs that they acclimated and regulated their internal body temperature to the vigorous exercise. In a 1977 paper, Drinkwater wrote that "these observations in addition to those reported earlier emphasize again that aerobic power must be considered as an independent variable when male and female responses to work in hot environments are compared." Her studies were instrumental; they proved there was no scientific basis for barring women from long-distance running, including the marathon.

More recently, exercise physiologist and nutrition scientist Stacy Sims disrupted the traditional exercise physiology and sports science model. An endurance athlete herself, Sims has long wondered whether the menstrual cycle played a role in athletic performance or if it was just a myth. When she rowed crew in college, she and her teammates often joked that they performed worse when they had their periods. But it was her experience competing at the 2002 Ironman World Championship in Kona, Hawaii, that made her realize that there's a huge gap in our understanding of female physiology.

Sims struggled in the heat and ended up in the medical tent after the race with hyponatremia, a condition where sodium levels in the blood are lower than normal, leading to swelling in the brain, which can be fatal. When she regrouped with her training partners, she found out that some women also had trouble with the heat and hydration, all of whom were about to start their periods. In contrast, the women who were at the beginning of their cycle raced well. It was curious, considering they all followed the same nutrition and heat-acclimation procedures. Sims wanted to get to the bottom of it.

When Sims returned to New Zealand, where she was working

toward a PhD at the time, she pivoted her research to focus on sex differences in sports. She's spent the past twenty years trying to understand the relationship between female hormones and athletic performance. In doing so, she's directly challenged the existing principles around exercise, nutrition, and health and their applicability to women, and her book *Roar: How to Match Your Food and Fitness to Your Female Physiology for Optimum Performance, Great Health, and a Strong, Lean Body for Life* has centered female physiology in conversations around sports science. With her rallying cry that "women are not small men," she has brought renewed attention to women's bodies and the need to help women understand and work with their physiology.

I n one sense, the men who were doctors and scientists were right: women's bodies are an important consideration when it comes to sports and physical activity—just not for any of the reasons they claimed. Women's bodies do differ from men's, primarily in relation to the reproductive system. But having a uterus, ovaries, breasts, and a wider pelvis doesn't make someone weaker or any less suited for physical activity, and it doesn't mean they shouldn't have the same opportunities as men. Those differences just make them different.

The distinct physiological and biological characteristics of female and male bodies can lead to variations in size, biomechanics, endurance, and patterns of strength and weakness. It doesn't necessarily mean one type of body is better than the other, but these attributes can fundamentally influence how the body feels and responds to exercise and a person's lived experience of sports.

For instance, men are, on average, bigger in stature compared to women. That means they have longer arms and legs, bigger organs, and greater blood volume. That also means that men have a higher capacity for aerobic work. Larger lungs mean they can extract a lot of oxygen and larger hearts mean they can circulate more oxygenated blood through-out the body. Muscle composition tells an interesting story too. While women and men have roughly the same percentage of type I (slow-twitch, endurance-type) and type II (fast-twitch, power-type) muscle fibers, the size of the fibers is different. In women, type I fibers tend to be bigger than type II fibers, whereas it's the opposite in men. It ex-plains why, when it comes to strength and power, men will generally perform better while women may do better in long endurance sports.

To understand our experience with exercise and sports, we need to become body literate. When it comes to anatomy, biology, and physi-ology, we have to start with an understanding of biological sex. Sex dif-ferentiation doesn't begin until somewhere between the first and second months of pregnancy. That's when the Y chromosome sends a signal to the embryo to develop testes and start producing testosterone, steering it down a stereotypically male pathway. In embryos without a Y chromo-some, the embryo instead develops ovaries and travels down a stereo-typically female pathway.

While the Y chromosome turns on the switch for sex differentiation, hormones are at the heart of the differences between female and male bodies. During puberty, these chemical messengers surge through the bloodstream and orchestrate changes throughout the body. People as-signed male at birth sprout up in height and begin to lay down more bone and muscle. In people assigned female at birth, hormones prompt the body to shape-shift and prepare for future childbearing. Hips widen, breasts form, and the body starts accumulating fat. The menstrual cycle begins, and female sex hormones like estrogen and progesterone start to ebb and flow.

Growing up, I assumed that the menstrual cycle was just a few days of bleeding. That's it. I didn't really piece together the "cycle" part of it. I didn't realize that it involved a rhythmic variation of hormones or that it was one of the most important biological rhythms—akin to the body's sleep-wake cycle, but instead of operating on a twenty-four-hour clock, the brain coordinates changes in the ovaries and uterus over roughly twenty-eight days (and two phases) to prepare the body for a potential pregnancy. Estrogen levels can increase fivefold over the course of a month, creating a vastly different physiological environment in female bodies compared to male bodies, where hormone levels remain relatively stable.

During the first phase of menstruation—the follicular phase—hormone levels are low. The body is busy shedding the endometrial lining of the uterus and bleeding starts. The brain, however, sends signals to the ovaries to prep a new batch of follicles (each one containing an egg) and produce more estrogen to get ready for ovulation, when an egg is released. Over the next fourteen days or so, follicles mature, the uterine lining rebuilds, and hormones like estrogen rise until they surge, triggering ovulation and the beginning of the second phase of the cycle—the luteal phase. Now, the body shifts into nesting mode to get ready for a possible pregnancy. The remnants of the follicle that released an egg begin producing progesterone and some estrogen, keeping hormone levels high. If a fertilized egg is implanted in the uterus, progesterone remains elevated. If not, hormones drop and the uterine lining starts to deteriorate, starting the period. Then the cycle begins again.

That's the textbook version of the menstrual cycle. Cycles don't always last twenty-eight days, nor are they evenly split between the follicular and luteal phases, with ovulation occurring on day fourteen. The reality is messier. Normal cycles can range anywhere between twenty-one and thirty-five days. Not only can cycle length vary between individuals, but a person's cycle can change month to month too: it may last twenty-six days one month and thirty days the next. Cycle and phase lengths also vary depending on age, stress, and lifestyle factors in addition to a person's race and ethnicity. They tend to be longer and more erratic during adolescence before settling down into a more predictable pattern during adulthood. Things start to get wonky again as a person approaches menopause.

While the main role of estrogen and progesterone is to regulate reproductive function, these hormones actually have a laundry list of jobs. For example, there are estrogen receptors throughout the body. Depending on where the hormone is active, it can influence different physiological systems—cardiovascular, gastrointestinal, musculoskeletal, and immune function—and even mood. Exercise physiology and nutrition professor Anthony Hackney found that some of these other jobs can influence how well you might be able to exercise, how good the exercise feels, and when you start to feel fatigued.

Hormones also spur breast development. The size, shape, and density of breasts depends on the composition of three types of tissue: connective tissue, glandular tissue (where the milk-producing glands are located), and fatty tissue. One reason we need external garments like bras and sports bras is because there aren't any bones or muscles to support breasts or control their movement. While we might joke about bouncing boobs, breast support is a big deal. One in four people with breasts says they are a barrier to physical activity, a ratio that increases to one in two among adolescent girls (and even higher among people with larger breasts). More than half of women (and over 60 percent of elite athletes) experience breast pain. A survey of 168 female runners at the 2012 London Marathon found that, all things being equal, each increase in cup size tacked on an additional four to eight minutes to a woman's marathon time.

Along with breast development during adolescence, the shape of the pelvis changes in people assigned female at birth. Estrogen softens connective tissues, causing the pelvis to broaden to prepare for a possible pregnancy and childbirth. The pelvis in those assigned male at birth, on the other hand, stays relatively narrow and grows concurrently with the rest of their skeleton. These structural changes can affect biomechanics, movement patterns, and, most critically, the integrity and function of the pelvic floor. This bowl of muscles, which lines the inside of the pelvis from the pubic bone to the tailbone, supports the pelvic, reproductive, and urinary organs like a hammock. It also acts like a trapdoor and helps control bladder and bowel function.

It wasn't until a few years ago that I learned that the pelvic floor does more than just hold organs up and inside. It's a key part of the body's core stability system, a set of four muscles—the pelvic floor, diaphragm, transverse abdominus, and multifidus—that regulate pressure inside the abdomen. Every time you breathe or move, the pressure inside the canister shifts, and the "core four" work in concert to adapt to these changes to keep pressure balanced. Balanced pressure is needed for everything from stabilizing your posture to transferring force so you can do things like lift something overhead or throw a ball. These muscles even anticipate motion, stabilizing your body to absorb ground reaction forces before you move or your heel hits the ground.

Over time, pressure on the pelvic floor can weaken the muscles and overload the system. Pregnancy and childbirth, constipation and straining, and even chronic coughing and sneezing can lead to dysfunction like urinary or fecal incontinence, prolapse (where the uterus, bladder, or bowel bulges into the vagina), pelvic pain, pain with intercourse, and even pain using tampons. While pelvic floor dysfunction can affect all humans, research has found a higher prevalence in women compared to men because women essentially have a wide, bottomless cavity punctuated with the vagina, urethra, and anus. "You've got more organs, more holes, more lifetime events [like pregnancy], so the possibility for dysfunction is going to be more prevalent in females," pelvic floor physical therapist Abby Bales told me, even for those who have never been pregnant or given birth. Other factors like genetics and hormone status can also influence the health and integrity of the pelvic floor.

A leaky bladder, prolapse, and pelvic pain aren't just inconvenient and embarrassing. They can affect a person's quality of life, their sports performance, and whether they choose to exercise at all. Active women and athletes in particular are susceptible to pelvic floor issues, especially if they take part in high-impact sports like running, rugby, and gymnastics, where the pelvic floor repeatedly absorbs high ground reaction forces, and strenuous weightlifting, where short bursts of exertion can cause pressure inside the abdomen to spike, making it more likely to overload the system. On average, more than one-third of active women and athletes report urinary incontinence—a number that's likely an underestimate. Adolescent girls aren't immune from these problems either.

While we often brush off incontinence or pain as the price of being a woman and playing sports, Rita Deering, assistant professor of physical therapy at Carroll University in Wisconsin, told me, "Women need to be educated about their bodies and know what is normal and what is not normal." And they need to know it's okay to seek help.

If we want to develop a scientific evidence base that's applicable to a more diverse range of bodies, we can't ignore factors related to biological sex. But here's the tricky thing—what I just told you about female and male bodies is a simplified version of a very complex story. The differences between female and male bodies aren't always cut-and-dried and entail more than just a pair of XX chromosomes or a pair of XY chromosomes.

Yet, most of us have been taught to think about sex as a binary attribute, largely determined by our genes. We've ascribed a tremendous amount of power and importance to this pair of so-called sex chromosomes as *the* determinant of biological sex. It turns out that the scientists who discovered these chromosomes at the turn of the twentieth century didn't want to call them sex chromosomes at all. The name "sex chromosomes" was meant as shorthand. It seemed like an oversimplification to think that a single pair of X's and Y's controlled just this one facet of human life and not recognize that these chromosomes had many other functions. While the Y chromosome does carry the SRY gene that sends an embryo down a typically male development path, it also contains many other genes. The X chromosome has more than a thousand genes, only 4 percent of which are related to sex and reproduction. But despite this, the name "sex chromosomes" took hold, along with more meaning than originally intended.

Biological sex involves more than just chromosomes and genes. When we separate bodies into categories like female and male, we tend to group them based on physical features like external genitalia (which is how doctors assign sex to babies at birth, rather than basing it on their chromosomes) and secondary sex characteristics like the presence of breasts, wider hips, and facial hair. But there are other features we can't see that also influence biology and physiology—internal genitalia (like the uterus and vagina or prostate gland and vas deferens), gonads (like the testes and ovaries), the types of hormones the body produces, and the body's ability or inability to respond to those hormones. These characteristics tend to cluster together. We associate a stereotypically female body with a vulva, uterus, vagina, ovaries, breasts, and estrogen production, but the story is more complicated than that.

It's becoming clearer that what's traditionally considered biological sex actually exists along a spectrum, and there's a lot of diversity in the actual physical expression of these characteristics. For each of the areas related to biological sex, the body can go down different paths. There isn't just one big fork in the road; there are many. People may end up with a collection of features that place their bodies squarely in a category traditionally thought of as "female" or "male." Or they may end up with variations in sex traits or chromosomes that fall outside these categories. People with intersex bodies (roughly 1 percent of the population) could have a number of combinations of X and Y chromosomes—

multiple X's, multiple Y's, and even a single X. They could have XY chromosomes but have external female genitalia. Others may produce high levels of testosterone, but their bodies may not respond to that hormone. There's a wide range of possibilities.

Biological sex isn't destiny, and it doesn't exist in a vacuum. It's very much entangled with gender—the socially constructed roles, norms, and identities that influence how we perceive ourselves and others. Gender and gendered norms greatly impact a person's experience of the world around them and their body in that space. The physical, social, and cultural environment shapes how girls and women learn to move (or not move) their bodies and the opportunities to be physically active available to them. It's tricky terrain to navigate, especially for nonbinary folks, transgender individuals, and intersex people who don't always align with the two sex-segregated categories that modern sports are based on. Because they exist outside the strict boundaries of what's considered "normal," their bodies are often used as a reason to exclude them from sports. There's no easy answer or solution, especially when confronting the norm can be uncomfortable and controversial.

Recognizing that sex (and gender) exist on a spectrum is crucial. We can't improve our understanding of exercise physiology and sports science if we don't study the full diversity of human beings. We can't make science more inclusive if we keep leaving out people because they don't fit into predetermined boxes. Thinking about sex as a continuum offers a new perspective on what we've always taken for granted.

As our knowledge of exercise physiology and sports science evolves along with our sociocultural context, change is brewing. Women are ready to acknowledge that their bodies aren't a nuisance or hindrance when it comes to physical activity and sports. Rather than ignoring it, people want to make informed decisions and understand how they can work with their physiology to be the best version of themselves. And for many, that starts with the menstrual cycle.

3.

Period Power

· • • ● ● ● • • ·

For years, doctors told Traci Carson it wasn't a big deal to skip her period. After all, she was an athlete. She played soccer for most of her life and ran track and cross-country in high school. Her senior year, she was even the kicker for her high school football team, the first girl to play on the varsity team in the school's history. Her doctors told her many fit and active people don't menstruate.

When Carson arrived on campus at the University of Michigan as a transfer student her sophomore year, she walked on to the crew team. But before she could officially join, she had to pass a physical. She gathered along with hundreds of other Wolverine athletes in what Carson described to me as a "cattle line" that snaked around the football stadium, nicknamed the "Big House." When it was her turn, a clinician asked Carson a few questions, including the date of her last menstrual cycle. When she responded, "Seven years ago," the clinician noted it in her records and then moved on. Carson was cleared to row.

"I knew at the time that it was probably not okay," Carson told me, about both her missing menstrual cycle and her underlying history of disordered eating. "But moments like that, where you're in an athletic

setting and they're telling you you're fine, you think, 'I must be fine.'"
Coupled with her previous doctors' lack of concern, she was led to believe nothing was wrong because she was an athlete.

Carson's wake-up call came after her annual physical at her hometown doctor's office. That day, there was a nurse practitioner seeing
patients—Carson thinks the nurse must have sensed that something
was wrong just by her disposition. "[She] asked me some very specific
questions and I just lost it emotionally. I broke down," she remembers.
Carson told the nurse about her absent menstrual cycle and eating disorder, and the nurse ordered some bloodwork. When the office called
with the lab results, Carson was told her "estrogen levels were menopausal," even though she was only a twenty-year-old college student.

It was a phrase Carson says she will never forget. She realized she
wasn't healthy. The science actually *said* she wasn't healthy. She wondered how this would affect her fertility, but she couldn't find any information on the odds of getting pregnant after not menstruating for so
long. Even years later, Carson says she's still grappling with her frustration over the lack of understanding among physicians and athletic staff
about the menstrual cycle—how it relates to women's health and well-
being, and how best to treat women, especially athletes.

When we first learn about our bodies—whether in health class,
in conversations with parents, during a section in biology class,
or in hushed whispers among friends—we learn that the menstrual
cycle is how the body prepares for pregnancy. However, clinicians acknowledge that it can tell them about much more than just fertility. It
can be a useful indicator of a person's overall health and should be considered a vital sign, much like body temperature, pulse, breathing rate,
and blood pressure. A normal, functioning menstrual cycle is a way of
saying a person has a healthy endocrine system. And that's a good thing.

The characteristics of a person's cycle—whether it's regular or miss-ing, light or heavy, painful or pain-free—offer a unique glimpse into the body's inner workings. Indications of irregularity can be an early warn-ing sign that something is amiss. "When we start to see the absence of the menstrual cycle or infrequent cycles, then those are signs that the hormones associated with reproduction are not functioning appropri-ately or perhaps at all," says Anthony Hackney, professor of exercise physiology and nutrition. "If they're not doing their primary job, they're not going to do their secondary jobs well either."

The menstrual cycle is regulated by a region of the brain known as the hypothalamus, which controls your body's production of hormones. When it's time for ovulation, the hypothalamus sets off a process to tell the ovaries to prepare an egg and make estrogen and progesterone.

Skipping a period could mean you're pregnant. Sometimes cycles just get finicky, especially in the first year or two of starting menstru-ation, before the body settles into a more consistent rhythm, and later in life right before menopause, when cycles cease altogether. Medica-tions and health conditions can also trigger physiological changes that throw off the finely tuned feedback loop that keeps reproductive hor-mones in balance. So can stress, anything from travel to inadequate sleep to a bad breakup. More often, for active people, a wonky or absent cycle is a sign that the body doesn't have enough energy to fuel the activities of daily life. Undereating, overexercising, or a combination of both can result in an energy deficit that disrupts their cycle.

When the body detects that it's under too much stress, it prioritizes the most important systems, like the nervous and respiratory systems, and shuts down nonessential functions, like reproduction. What was once a well-choreographed dance of hormones is now going haywire. The erratic pattern can lead to irregularities like anovulatory cycles (when a person has a cycle without ovulating), extra-long cycles, a shortened luteal phase (a sign of low levels of the hormone proges-terone), or no periods, but there aren't always clear-cut symptoms indi-cating that something is wrong. These disruptions may last a short period of time, until internal systems rebound, or they may fundamentally alter your baseline, with menstrual cycle dysfunction and rejiggered hormone levels becoming the new normal. A menstrual cycle that plays hide-and-seek can be a cause for concern when a person hasn't started their period by age fifteen or doesn't menstruate for three or more consecutive months after previously having a period—both a form of amenorrhea—or when a person has consistent cycles longer than thirty-five days.

On the surface, an absent period may seem like it's not a big deal—it may even feel like a blessing. Who doesn't wish away their cycle and the nuisance of buying period products? But when the body doesn't experience the consistent monthly surge of hormones, it can lead to a cascade of problems. Estrogen plays a key role in building bones; it tips the balance toward bone formation over bone breakdown and resorp-tion. Without the regular flood of estrogen, especially during adoles-cence and young adulthood, girls and women may wind up with weaker bones, leaving them at greater risk for stress fractures and early onset osteoporosis. Studies also suggest that menstrual cycle disturbances may be linked to changes to the cardiovascular system, which can con-tribute to health problems like high blood pressure and coronary artery disease.

The lack of estrogen can impair training adaptation too. When thir-teen young elite runners were followed for a year, the eight athletes who didn't menstruate spent more days injured and ran less total mileage compared to their counterparts who had a normal cycle. Only those ath-letes with a regular period saw improvements in their performance.

Doctors oftentimes recommend hormonal contraception to people who aren't routinely menstruating as a way to jump-start and manage their cycle. Oral contraceptive pills, arm implants, shots, patches, vaginal rings, and some intrauterine devices (IUDs) introduce synthetic versions of female sex hormones into the body that then suppress and override the body's natural hormonal rhythms. (The exact type and dosage of hormones varies depending on the specific type of contraceptive.) Consider the oral contraceptive pill, one of the most popular forms of contraception. It's designed to imitate the typical twenty-eight-day menstrual cycle by delivering a consistent supply of estrogen and progestin (a manufactured form of progesterone) with twenty-one days of active pills, followed by seven days of placebo pills that contain no hormones.

The pill seems to solve the problem of menstrual dysfunction: it regulates the cycle, provides an influx of missing hormones, and induces a cycle (in most cases). However, the hormonal profile of a person on the pill, with its three-week steady state, is vastly different from the hormonal peaks and valleys experienced by a naturally cycling person. And the bleeding that occurs during the placebo week isn't a true period. Instead, it's the body's response to the withdrawal of external hormones. What's more, because everything seems to work like clockwork, hormonal contraception can mask the presence of an underlying problem but doesn't solve it. If the root cause of the disturbance isn't addressed, it will still be there when hormonal contraceptives are discontinued.

Many people, even those with normal menstrual cycles, choose to use hormonal contraceptives for reasons beyond avoiding pregnancy. This form of birth control offers a way to manage periods and cycle-related symptoms like cramps, bloating, and heavy menstrual bleeding. For people who are active or competing in sports, it can provide some semblance of control, even a way to manipulate the timing of one's period. It's no wonder that roughly half of elite athletes report using hormonal contraceptives. They don't have to worry if they'll start bleeding on race day or in the middle of a major tournament, and they don't have to fret about leaking through their clothes, especially if their uniforms include white shorts.

However, these distinct reproductive hormonal profiles—between someone who is cycling naturally and someone on hormonal contraceptives—matter. External hormones may influence a person's physiology in ways that are different from naturally occurring hormones, including how the body adapts to and recovers from exercise. While research on the effect of hormonal contraceptives on exercise performance is limited, some studies in elite athletes suggest they may be linked to higher levels of inflammation, which could mean reduced sports performance and recovery. Hormonal contraceptives may also influence the way the body regulates temperature, changing how blood flows to the skin and the threshold at which the body starts to sweat. These alterations can impact how the body feels and responds to exercise in the heat.

While the period is a telltale sign of the menstrual cycle, it's not the only aspect that can cause major headaches. Most menstruating people acknowledge that the cycle's accompanying symptoms can greatly affect how they feel while active. A 2020 study of more than 6,800 users of the fitness app Strava from around the world found that more than 80 percent of exercising people who menstruate frequently experienced at least one cycle-related symptom. The most common complaints were mood changes, anxiety, fatigue, abdominal cramps, headaches, increased cravings or appetite, and breast pain or tenderness. The more symptoms a person experienced, the more likely they were to change their training program or even skip a workout session or sporting event.

The trend persists among high-level athletes too. Professional tennis player Danielle Collins experienced years of unbearable periods. Abdominal and pelvic cramps. Sciatica-like nerve pain. Sharp pain in her spine. Anti-inflammatories and rest didn't alleviate her symptoms, and doctors told her that painful cycles were normal. But Collins couldn't practice or train consistently, inhibiting her ability to move up in the rankings. In early 2021, she pulled out of the Adelaide International tournament because of severe back pain and collapsed on court during the Australian Open. A few months later, she had emergency surgery. It turned out that Collins has endometriosis, a condition where uterine-like tissue grows outside the uterus. In an interview with the Women's Tennis Association, she said, "The agony that I experienced from my menstrual cycles and from the endometriosis is some of the worst pain I've ever had." More than 70 percent of elite athletes noted side effects linked to their cycles, like reduced fitness. Yet even though these symptoms can be debilitating and a source of anxiety to the point of changing a person's exercise habits, menstrual cycles aren't an open topic of conversation. No one talks about it.

It's an interesting dichotomy. Athletes, especially at the professional, Olympic, and Paralympic level, often investigate every possible avenue to eke out a competitive advantage. Along with their coaches and support team, they'll discuss matters of pre- and post-workout nutrition, biomechanics, achy muscles and joints, bowel movements, and even factors like the firmness of their pillows. But mention the menstrual cycle and it's like shooting a flare in the sky, warning people to change the topic. It can be especially awkward to discuss female physiology-related issues with athletic and medical staff who are men and don't fully appreciate the lived experience of these hormonal changes.

It's in part because people who menstruate often don't understand what's going on inside their own bodies—the basics of the cycle, the hormones that orchestrate it, and how it could affect the body outside of reproduction. For most people, it's either you're on your period or you're not. That's it. It's not surprising when you consider that 72 percent of fitness buffs and avid athletes reported not receiving any education about their menstrual cycle and its relationship to exercise.

Elite athletes are also left with gaping holes in understanding their own bodies. Growing up, two-time X Games gold medalist Mariah Duran was told that one week out of the month would be rough because of her period. But when Duran started learning more about her cycle and paying attention to her body, she said her mind was blown. She never knew there were different phases to her cycle or that it could affect how her body functions and feels when she's skating. "For the longest time, I thought [the menstrual cycle] was a negative thing when in reality, your body is just doing what it has to do," she recounted during an Instagram Live session.

"It drives me mad that we are not having these conversations with women who are using their bodies as their instrument," says physiologist Emma Ross, who developed the SmartHER program at the English Institute of Sport and founded the Well HQ to empower women to optimize their health and performance. "They haven't been given the permission to diligently explore strategies to make those symptoms go away or alleviate them in a way that makes them perform better." As a result of the tight-lipped culture, Ross believes that people who menstruate are scared to say they feel terrible on certain days of the month. They're worried they'll be judged as weak, and that it will be used against them as a reason to not promote them to the next level or to the A team.

But Ross says it's not just about elite athletes. The silence trickles down to active women and girls too. She sees far too many young people drop out of sports because of preventable reasons, like fear of judgment or fear that their periods will get in the way. "We haven't created this safe space where we give women a voice to say it's not all weakness.

You're just trying to figure this stuff out," she says. And doctors, coaches, and athletic staff need to be educated about women's health and physiology too so they can better support active women and athletes.

While period-related symptoms were first described in a scientific study in 1931 by gynecologist Robert Frank (who originally called it "premenstrual tension"), doctors and researchers still don't know what causes premenstrual syndrome (PMS). Even though 90 percent of people who menstruate experience at least one PMS symptom, there's little research on the topic, leaving women and their physicians with limited tools to treat the physical and mental discomfort. When there are no viable treatment options beyond ibuprofen and a heating pad, discussing menstrual cycles can feel like a futile conversation. So why bother?

In recent years, the taboo around the menstrual cycle has started to crack as athletes have spoken more openly about their personal experiences. At the 2016 Rio Olympics, periods were part of the pool deck conversation following the women's 4×100-meter medley relay. After finishing fourth, the team from China spoke with a Chinese journalist. Fu Yuanhui squatted down out of sight as her teammates were interviewed. Fu had won the bronze medal in the 100-meter backstroke earlier in the meet and was expected to help put China on the podium in the relay. But her swim was unremarkable. When it was her turn to address the cameras, her teammates helped her stand and she visibly grimaced. Between heavy breaths, Fu said that she felt she had let her teammates down. When asked if she was experiencing stomach pain, she explained that she had started her period the previous day and felt particularly tired. "But this isn't an excuse. I still didn't swim well enough," she said.

In the past, athletes would have glossed over the real reason their performance wasn't up to par, but Fu spoke matter-of-factly. Fans praised her for breaking the silence around something faced by people who menstruate every month. Her story encouraged retired Japanese swimmer Hanae Ito, who competed in the 2008 and 2012 Olympics, to open up too. In 2017, Ito shared that she was on her period while competing in Beijing, writing, "I'm not going to say I would've won a gold medal if I knew then that there were options to deal with period-related weight gain, acne and other issues I was having, but I definitely think I would've performed better."

More recently, the day after the women's marathon at the Tokyo Olympics, Lonah Chemtai Salpeter took to Facebook to explain why the wheels came off late in her race. Salpeter, who holds numerous Israeli national records, including in the marathon, came into the event a potential contender. In brutally hot conditions, she ran with the lead pack and was in bronze medal position. But with 4 kilometers to go, her cramps were so bad that she had to stop and walk. She finished in sixty-sixth place. On any other day, she believed she would have won a medal. She wrote, "Knowing that a lot of female professional athletes face health issues, I feel that I have to play an instrumental role to help and break taboos around the female period."

These conversations have helped normalize discussions about hormones and have spurred interest in menstrual cycle tracking. Period tracking isn't new—women have kept tabs on their cycles formally and informally for years, mostly to anticipate ovulation and a potential fertile window—but the practice only recently made its way into the arena of sports and exercise. For example, during the 2019 season, members of the Charlton Athletic Women's Football Club in England monitored their cycle along with wellness data, something that only 40 percent of players did prior to the start of the season. By the end of the season, 90 percent of players reported that monitoring made them more conscious of their individual cycle, rhythms, and symptoms. Regular tracking also made it easier to broach seemingly off-limits topics within the club. Players felt more comfortable discussing female-related issues with teammates and their strength and conditioning coach. It helped the medical and support teams identify underlying health issues in some athletes, such as absent, irregular, or abnormal periods.

Apps like Clue and Flo make the process easier too. Women can log exercise activity and energy levels in addition to conventional menstrual cycle symptoms. Wearables from Garmin, Fitbit, WHOOP, and Apple incorporate period tracking functions too, giving people the ability to note their cycles and symptoms next to their runs, bikes, walks, and swims. These apps are part of the growing "femtech" market—products, services, and technology solutions designed to empower women to take control over their health and to improve women's wellness. And they're popular. Globally, an estimated fifty million people use menstrual apps.

Can period trackers tell you more than just when to expect your next period? Increasingly, coaches and athletic staff want to learn more about hormones and what they need to keep in mind when working with athletes who menstruate. Should people who menstruate train differently from those who don't? Do fueling needs and training adaptations vary depending on whether it's the high- or low-hormone phase of the cycle? Should training be periodized to take advantage of the naturally occurring surges of hormones? Could doing so help improve fitness and performance? Would it prevent injury? These aren't far-fetched questions, considering that reproductive hormones do influence an array of biological processes such as how the body regulates heart rate, blood volume, and temperature; how it dissipates heat and balances fluid; how it grows and repairs muscles; and how it metabolizes food.

While scientists have studied female reproductive hormones for decades, they've taken a greater interest in the role of hormones on exercise adaptation and performance in the last ten years. And they're uncovering some interesting trends. For instance, research suggests that estrogen may boost muscle gain and could be the reason why muscle strength seems to vary across the cycle. In one study, twenty naturally menstruating women took part in a four-day-a-week leg strengthening program. They trained one leg during the first half of the cycle (follicular phase) and the other leg during the second half (luteal phase). After three menstrual cycles, researchers saw greater increase in strength and muscle diameter in the leg trained during the follicular phase. That might mean people could gain more muscle if they concentrated their lifting sessions during the earlier weeks of the cycle.

Meanwhile, exercise might seem different during the luteal phase. During the high-hormone phase, the body conserves carbohydrates. Without this readily available source of energy, some types of exercise can *feel* harder. Plus, as the body moves from the follicular to luteal phase, higher hormone levels appear to inhibit the body's ability to bounce back from workouts. Studies have found that metrics like heart rate variability (the fluctuation between heartbeats) decreases while resting heart rate increases, all signs that indicate the body isn't recovering easily. These physiological changes can also make people feel flat, causing a dip in motivation and desire to train at all.

Several companies have seized on the opportunity to try to make sense of the emerging research and translate it into concrete advice. Orreco, a tech company that blends sports science with data-driven technologies to improve performance, introduced its app FitrWoman in 2016. People can log traditional symptoms like blood flow, cramps, bloating, and tender breasts next to more sports-specific concerns like heavy legs, injury, and increased breathing rate. They can record physical activity or sync their data with Strava. Based on a person's current menstrual cycle phase, FitrWoman offers in-app education about what's happening in the body, as well as suggestions for working with the body's physiological and hormonal peaks and valleys. Wild AI takes it a step further. It uses artificial intelligence to comb through an individual's specific data and provide personalized training, nutrition, and recovery recommendations best suited to each menstrual cycle phase.

The rush to prescribe exercise strategies based on the menstrual cycle is understandable. Historically, people who menstruate have been overlooked when it comes to sports science. Now that scientists are finally paying more attention, people want clear answers. Anecdotally, some say that syncing their workouts to their cycle works. Many report results from focusing on high-intensity and strength training during the follicular phase and sticking with slower-burn endurance workouts, mobility, and restorative sessions like yoga during the luteal phase. Social media influencers have posted about their experiences and published guides so others can follow suit. Fitness coaches are designing programs for clients. Nike and Peloton offer advice on how to tailor your training to your cycle (with the help of their apps, of course). The media jumped on the story too, implying that period tracking can give women an unparalleled edge by unlocking previously unknown potential. It's the epitome of personalized training.

When I first came across the idea of cycle syncing for exercise, I was excited about the potential to make sports science guidelines more applicable for menstruating people in a way we haven't been able to before. But at the same time, I worried that the hype was getting ahead of the science, especially since much of the research is still preliminary. I was afraid that in the rush to draw attention to women's physiology, we were settling on a simple fix by placing all bets on the menstrual cycle being *the* answer to everything.

While tracking hormonal cycles can provide a wealth of information about one's own body, it's not a crystal ball. Given the current evidence available, simply determining where you are in your cycle won't reveal a road map for what type of training to do and when. Emma Ross says we have a few nuggets of information, but researchers still know little about the magnitude and direction of these possible effects of the menstrual cycle. In other words, more studies are needed to understand what it all means.

A team of researchers in the United Kingdom tried to make sense of the available evidence in a 2020 systematic review and meta-analysis because "at present, data are conflicting, with no consensus on whether exercise performance is affected by [menstrual cycle] phase." They gathered seventy-eight previously published studies that investigated the relationship between exercise performance and the menstrual cycle among naturally cycling people. The studies looked at a variety of strength and endurance indicators, compared them to different points in the cycle, and looked for patterns in the data. The most apparent trend emerged when the team compared outcomes from the low-hormone, early follicular phase to all other phases of the cycle. They found a slight decrease in performance.

Yet, when they took a step back and looked at the studies as a whole, the results essentially canceled one another out. For every study that showed that women performed worse during the early follicular phase, there was another study that demonstrated the opposite. The study authors wrote, "The implications of these findings are likely to be so small as to be meaningless for most of the population." (Some critics point out that the study only compared outcomes that measured performance at a point in time and didn't look at training adaptations that may emerge over a longer period.)

The same team also examined the effect of the oral contraceptive pill on exercise performance. They similarly extracted and analyzed the data from previously published studies. When they compared exercise performance of those who used oral contraceptives to those with a natural menstrual cycle, they found that oral contraceptive users experienced slightly worse performance. But once again, the overall effect was trivial at best.

Scientists say a key reason it's so hard to make sense of the research findings is the lack of agreed-upon standards within the field. If there isn't consensus on the definitions for menstrual cycle phases, criteria for selecting participants, and study methodologies, it's difficult to compare studies head-to-head. Some researchers may not verify menstrual cycle status and phase with blood tests, relying on participants to self-report. Or studies may not segregate naturally cycling participants from those who use hormonal contraceptives, lumping them into a single experimental group despite the fact that their hormonal profiles are not the same. It calls into question the overall trustworthiness of the data. If the researchers aren't 100 percent confident that participants were in the cycle phase they're investigating, how reliable are the findings? If researchers aren't concerned about conflating the effects of external hormones with naturally occurring ones, are they drawing accurate conclusions?

What everyone can agree on is that we need more studies. It may take years to develop the foundation of quality research needed to determine whether to modulate exercise across the different phases of the menstrual cycle and how to do so. Even then, some experts are skeptical that we'll arrive at a comprehensive, one-size-fits-all blueprint. Kirsty Elliott-Sale, female endocrinology and exercise physiology professor and the lead author of the meta-analysis on oral contraceptives and exercise performance, told me that this isn't an area like nutrition where there's often a clear and simple mathematical formula—you weigh this much, so you should eat this many grams of carbohydrates. Translating research findings isn't as simple as "estrogen does X and progesterone does Y." Sometimes the two hormones work together, and sometimes they work against each other. Their influence depends not only on their individual concentration levels but the ratio between them, which constantly shifts throughout the cycle and lifespan.

To complicate matters further, studies report on the *group* response to an experiment, essentially the average effect across all participants.

But individuals don't all respond to hormones to the same degree, nor do they always respond the same way each month. The lived experience of people may be completely different, even if they are the same age, in the same menstrual cycle phase, or on the same day of their cycle. Some are less sensitive to the rise and fall of hormones and experience little disturbance to their physical and mental state over the course of the month. For others, it feels like they're riding out a storm in a small boat, bobbing up and down along with the big recurring shifts over their cycle. One person might experience tremendous progression when syncing their training to their cycle, while another might not see or feel any difference.

For a significant number of people, cycle-based recommendations aren't relevant because they either don't have a cycle, experience menstrual irregularities, or take hormonal contraceptives. "You've got up to thirty percent of athletes with menstrual dysfunction. You've got fifty percent of athletes on hormonal contraception. It doesn't really leave many athletes where you're going to get this right," says Clare Minahan, associate professor in sports science at Griffith University in Australia. Instead, Minahan believes it's critical to focus on a person's overall health and well-being first, like making sure they even have a regular cycle, before attempting to fine-tune a cycle-based approach to training and nutrition. These bigger-ticket items are the necessary foundation to support fitness, performance, and athletic success over the long term. "Once we get that right, then we can start to optimize performance," Minahan says.

When people have a regular menstrual cycle and are more aware of what's going on in their bodies, they're better equipped to notice individual patterns—and subsequently when something's off—and can home in on the peaks and valleys of their own physiology. With that data in hand, they can then make informed decisions about exercise, nutrition, and hormonal contraceptives.

More than 57,000 fans packed into the Stade de Lyon in France to watch the U.S. Women's National Team (USWNT) battle the Netherlands during the 2019 Women's World Cup final. It was a nail-biting, scoreless first half as the Netherlands repeatedly thwarted the Americans and their attacks on goal. But Jill Ellis, then head coach of the USWNT, knew it was only a matter of time. The game had to break one way or the other.

Shortly after the teams took the field for the second half, the break Ellis was waiting for materialized. The United States was awarded a penalty kick after Dutch defender Stefanie van der Gragt was called for a high boot in the penalty box. Co-captain Megan Rapinoe lined up her shot, tucked her purple hair behind her ear, and exhaled before sliding the ball into the lower right side of the goal. The fans erupted. Rapinoe trotted to the corner of the field and struck her iconic pose in her brilliant white uniform—her arms lifted out wide on a diagonal—before Alex Morgan and the rest of the team swallowed her up in celebration. Less than ten minutes later, Rose Lavelle dribbled up the open field, split the defenders, and sliced a kick off her left foot, sending the ball into the goal. The USWNT was up 2–0. When the final whistle blew, it was official. The USWNT earned its historic fourth star, symbolizing the team's four World Cup wins.

After the confetti fell and the trophies were distributed, news outlets began the postmortem on the team's performance, dissecting its plays and looking for insight into its competitive edge. And judging from the headlines that followed the final, the secret to the USWNT's stellar title defense was simple and seemingly unconventional. Every member of the team tracked their menstrual cycle in the lead-up to the tournament. On the surface, the USWNT story seemed like more hype for period-based training, but a deeper look reveals that their approach was more nuanced than that. Their experience offers a potential path forward for how to think about the menstrual cycle in relation to athletic performance.

The strategy was the brainchild of Dawn Scott, the high-performance coach for USWNT and the National Women's Soccer League (NWSL) at the time. Scott studied physics at university in the 1990s, which might explain her methodical and logical approach to sports science. For her, elite sports is all about the margins. She constantly searched for factors that could influence an athlete's performance, no matter how small. This included obvious areas like training, strength, and mobility, but also recovery, hydration, nutrition, and sleep. Tweaks across these areas could result in tangible improvement for a player—a percentage point here, a percentage point there—that, when taken together, would result in a stronger, fitter, and more resilient athlete over the long term.

When Scott joined U.S. Soccer in 2010, she oversaw the buildup of a robust data tracking and monitoring system to support her quest for incremental improvement. Initially, she relied on traditional fitness evaluations like movement screens and the grueling beep test, where players must run a set distance before the sound of a beep. Soon, players began donning heart rate monitors around their chests and GPS trackers tucked into their sports bras. She asked players for daily wellness reports on off-pitch factors too, like mood, quantity and quality of sleep, fatigue, soreness, nutrition, and hydration. Slowly, Scott started to piece together a fuller picture of each athlete and their norms, which she used to spot opportunities for improvement and pinpoint areas where a player may need additional support.

Still, Scott knew there was one area she was neglecting—the menstrual cycle. From her own experience, she was aware that issues like cramps could be debilitating, forcing players to underperform on the field or, at worst, miss practice or a game. And she knew that the national team players didn't have any real strategies to manage their symptoms. There was a stark contrast between the way injuries were handled versus cycle-related symptoms. If an athlete experienced hip pain, a physician would conduct a physical exam and the athlete would undergo diagnostic testing like an X-ray or MRI to assess what, if anything, was damaged or contributing to the pain. They then determined the best course of treatment and rehabilitation. If one strategy didn't work, they considered another until symptoms dissipated and the hip healed.

Scott thought that if she could mitigate cycle-related symptoms, she could help the team show up as the best version of themselves on any given day. In 2017, she asked athletes to log the start date of their period. By the end of 2018, Scott noticed some interesting relationships between the player wellness data she already collected and the menstrual cycle data. She started to see some patterns: a couple of players who were fatigued, sore, or didn't sleep well a few days before their period; other players reported certain symptoms around the same point in their cycle every month.

Scott wanted to dig deeper but knew national team members weren't comfortable talking about menstruation. If she wanted players to report on their monthly cycle and describe what they were experiencing, she had to help them understand the relationship between female physiology and performance. She had to convince them that fumbling through awkward conversations would be worth it in the long run.

In 2019, she and another member of U.S. Soccer's sports science team traveled to the nine NWSL pro teams at the time and delivered educational workshops to players and coaching staff. Even before getting to the heart of the conversation, some players slid down in their chairs and pulled their hoods over their heads, cinching the openings tight around their faces. During one presentation, the men left the room under the impression that they weren't needed since the session would cover "women's things." But Scott asked the staff members to return and participate. "For me, that's part of it—making the environment and topic comfortable for players to talk openly with male staff as much as with female staff," Scott explained to me. During the hour, the conversation became less about menstrual cycles and periods per se and more about how reducing symptoms and improving recovery can improve performance on the field. Scott says that's when players started to get it.

Scott then started working with Georgie Bruinvels, senior sports scientist at Orreco and the brains behind the company's FitrWoman app. After the educational workshops, national team players started using FitrWoman to track their cycles while also continuing to submit daily wellness reports. Bruinvels also consulted with each player individually. Soon, Scott and Bruinvels not only had cycle start data for each athlete, they also had the average cycle length and reported symptoms. They noted repeat symptoms, severity, and when they occurred. Scott monitored the current menstrual cycle phase and symptoms for all athletes on an integrated dashboard, a system that was far superior to twenty-plus individual menstrual diaries.

With this information in hand, Scott and Bruinvels used the patterns as signals for when to ramp up activities or take things down a notch so players would feel good across the entire cycle. Over time, Bruinvels built individual symptom management plans and a spreadsheet detailing the timing of every player's menstrual cycle in the lead-up to the 2019 World Cup tournament. Then she alerted Scott when players moved into a phase of the cycle where they were more likely to experience symptoms. For example, if an athlete showed decreased signs of recovery during a particular phase, Scott would remind them to keep an eye on their post-workout strategy. If a player was prone to menstrual cramps, Scott promoted more anti-inflammatory foods with that athlete at specific times. "It was a win-win. We did see reduction of symptoms in players," Scott told me.

Scott says period tracking and symptom management was just one of probably twenty to thirty things the team implemented that could have contributed to their success in France. Importantly, it taught players to pay attention to their bodies and notice when they felt good and when they felt off so they could make informed decisions about how to combat their symptoms. The actual adjustments Scott and the USWNT implemented were purely related to nutrition, recovery, hydration, and sleep—things athletes should be doing anyway to promote optimal training adaptation and recovery. They didn't use the menstrual cycle as a limiting factor to say a player couldn't participate in a certain type of training because of their cycle phase, and they didn't program players' training around their cycles, which was easy to assume from the stories in the media. "We don't know enough yet in terms of what training might be beneficial or optimal at any stage of the menstrual cycle," Scott says.

What was revolutionary about the USWNT's approach was that they treated the menstrual cycle as what it is—another piece of the physiology puzzle. It goes back to Scott's detailed player profiling and monitoring system. Each player responds to training differently. Each player has their own history of injury as well as unique anatomical and biomechanical areas of strength and weakness. Each player has their own lived experience of their menstrual cycle. Since the menstrual cycle has real physical, mental, and emotional impacts on players, it should be considered as one part of the larger matrix to make sure each athlete is ready for game day. And when players are in sync with their bodies, they are more confident in their training, nutrition, and recovery, all of which can eliminate a lot of anxiety.

As USWNT defender Kelley O'Hara told me, "It's our physiology."

As we wait for more scientific evidence to emerge, people can begin by engaging in "me-search." The current research offers signposts to indicate what may be happening in the body. But to make sense of this information—and whether it's applicable to you—you need to track your cycle and note your individual patterns. Once you're aware of your own rhythms, you can start testing bits and pieces of the research. Some strategies may work. Some won't. And this data is meaningless if you're not paying attention to your body in the first place.

While there may not be one blueprint to guide athletes and coaches on how best to exercise and train based on the menstrual cycle, a blueprint may not be the end goal. There's a cultural shift underway. The conversations—among scientists, people who menstruate, coaches, and athletic staff—and calls for more (and better) research are drawing greater attention to this topic. They're empowering more people to understand their own bodies and to recognize that menstrual cycles play an integral role in overall health and well-being. As the spotlight on the menstrual cycle intensifies and expands, it's casting aside shadows and myths in other areas of sports science too, the most vital of which is nutrition.

4.

Fast Fuel

· ◦ ◦ ◉ ◆ ◉ ◦ ◦ ·

In the years following the 2016 Summer Olympics in Rio, Rowing New Zealand, the country's governing body for the sport, uncovered some startling information. Most of their top elite women rowers didn't have a regular menstrual cycle and they weren't adequately fueling their bodies for their high level of physical activity. In other words, no one was eating enough food.

For Christel Dunshea-Mooij, senior performance nutritionist for High Performance Sport NZ, the findings set off alarm bells. Humans need a certain amount of food to power the body's activities, from basic life-sustaining functions like breathing, digestion, and circulation to daily movement. If a person is physically active, their body demands more energy to fuel its increased activity. Nutritional needs go up, both macronutrients like carbohydrates, protein, and fat, as well as micronutrients like vitamins and minerals. Athletes typically have even higher calorie and nutrition requirements, not only to keep up with the demands of their training but also to adapt to exercise and prevent illness and injury. If they don't eat enough, they won't adapt, recover, or perform well.

Dunshea-Mooij knew that Rowing New Zealand needed to do something. She took the results to the coaches and together they sent a clear message to all their athletes: Eat more.

Jackie Kiddle was one of the athletes found to be in the danger zone for underfueling. She told me it took a little convincing to get her on board with the new mindset. As a lightweight rower, undereating was a customary practice; it was what athletes did to make weight. But Dunshea-Mooij and the high performance team flipped the script on how they discussed food and weight with the athletes. Instead of using rhetoric like "lean is fast" and "ideal race weight," both common parlance in the sport, the team emphasized the benefits that came along with eating well—how nutrition and physiology directly connected with rowing, the sport the athletes loved and cared about, and how taking care of these factors could help them achieve their goals.

Once Kiddle committed to fueling better, it was a revelation. If she ate more, she could train more and work harder on the water without fading at the end of training sessions as she used to. If she trained more, she'd get faster and stronger. Plus, she'd stay at the same weight. It blew Kiddle's mind and she knew she couldn't go back to her old habits. She and her teammates pledged to eat plenty and eat well (and even call one another out when they didn't). Together, the athletes, coaches, and sports science staff changed the culture around diet within the women's program, shifting their approach to emphasize long-term health and performance over the specifics of weight and body fat percentage.

Across Rowing New Zealand, athletes became stronger, faster, and healthier. At the 2019 World Rowing Championships, Kiddle and her lightweight double sculls partner Zoe McBride won gold. At the 2020 Tokyo Olympics, New Zealand's women's boats won medals in four of the seven women's events, making rowing one of the country's most successful sports. "[It's] a direct example of what happens when you start fueling properly," Kiddle says. "The results spoke for themselves. It was incredible."

Rowing New Zealand's rallying cry to "eat more" isn't often heard around women's sports. Yet women have long struggled to match their nutritional needs to the requirements of their physical activity, regardless of the sport or whether it's at the recreational, club, high school, or elite level. A recent study found that professional women and men soccer players need similar amounts of energy each day, proportional to each person's lean body weight. However, when researchers looked at the number of calories the women athletes consumed and expended over a twelve-day training camp, they found that 88 percent did not meet their daily energy needs.

Across multiple arenas, women athletes are being asked to push the boundaries of performance in sports. But if women are starting with a half-empty fuel tank, they can't execute to the best of their abilities or ensure their long-term health. Is nutrition the stumbling block that's holding women athletes back?

When it comes to nutrition, the reasons why women falter can be simple and unintentional. Some people just can't keep up with the increased need for fuel that accompanies a step-up in training volume or intensity—whether they are just starting to work out or are jumping to the next level of their sport. They can inadvertently dig themselves into an energy deficit because they eat the same amount they always have while expending more calories than normal. Other factors can be external. When work, life, and family obligations pile up, nutrition can fall by the wayside. Since women tend to shoulder most of the household and caregiving responsibilities, they are more likely to be forced to care for their own needs around the margins, if at all. They may neglect eating before, during, and after workouts because there just isn't time. Good food can be expensive, too, and may not fit into everyone's budget or may not be accessible in every neighborhood.

For women athletes, there's also often a fraught relationship between food, sports, and body image that complicates matters. While body composition can play a role in some sports, it's not the final arbiter of performance. Yet the belief that a slim, lean physique is a competitive advantage pervades in endurance sports (like rowing, running, and cycling), weight-class sports (like wrestling or judo), aesthetic sports (like dance, figure skating, and gymnastics), and fitness writ large. Add in society's obsession with diet culture and its accompanying anti-fat bias, and there's tremendous pressure to conform to a specific idea of what an athlete should look like rather than asking whether they're eating enough and often enough.

These beliefs feed into a sporting culture that can turn a blind eye to, and at times promote, chronic dieting and disordered eating behaviors, sometimes without even realizing it. As an up-and-coming athlete, professional rock climber Beth Rodden benefited from her petite size, a body type that can be advantageous in the sport. She shimmied along hairline-thin fractures in giant granite walls, slotting her fingers into cracks and gripping the faintest footholds with the rubber of her shoes on her way to establishing routes that other climbers thought were impossible. She won competitions. She earned sponsorships. She was featured in movies and advertisements. Her accomplishments validated her propensity to restrict calories and underfuel, which only seemed to lead to more accolades. It created a vicious cycle that put her at risk, and by her late twenties, Rodden's body broke down. "Losing weight worked for my short-term performance gains but was extremely harmful in the long run," she writes in an essay for *Outside*. "Tendons, ligaments, bones—they all started to collapse after 15 years of deprivation. My climbing cascaded from elite to elementary in a matter of months."

As a swimmer at the University of Florida in the mid-2000s,

Olympian Caroline Burckle kept getting mixed signals from coaches. "Am I supposed to be strong, skinny, pretty, muscular, or fast?" she recalls thinking during her sophomore year. It was disorienting for Burckle because she'd always prided herself on being in tune with her body. It was her superpower. But now, she didn't know what to think. When she walked out on the pool deck, people commented that she looked great, yet intuitively, she knew she was too thin and overtrained. After a breakout freshman year, when Burckle set numerous records and came in second in the NCAA championship meet, she ramped up her training. A lot. Her body wasn't recovering from the endless laps, dryland training, and weightlifting, and she didn't get a lot of guidance on how nutrition factored into the equation. In her words, she tanked sophomore year. She entered the NCAA championships seeded first in her events but finished last in her heats.

Part of the problem is that navigating the relationship between sports and nutrition can be confusing, especially since active and athletic girls and women often aren't taught what it means to nourish their bodies or how to do it. It's assumed you know how to eat because you have to eat.

It's not something I learned about in high school or college—the tidbits I picked up were mostly gleaned from magazine articles. Seeking help from a nutritionist didn't seem feasible. It was a luxury, an expense that wasn't covered by health insurance. And it was also considered an implicit admission that you had an eating disorder, or some other problem that needed to be fixed. It was only in my late twenties and early thirties that I started to seek out more resources on nutrition. Even then, I didn't realize that the guidelines wouldn't necessarily help, since the underlying science was largely based on research conducted with young, fit men and paid little attention to sex-based differences.

When women follow general sports nutrition advice, they may or may not experience the same benefits and results reported in the studies or among men. The reason could be as simple as differences in body size and composition. Since women are typically smaller than men and have less lean body mass, recommendations may need to be scaled or adjusted. Or it could be the influence of the different hormonal environments in women's and men's bodies, which can affect factors like metabolism, fluid retention, recovery, and performance.

Even now, the field of nutrition, as it pertains to women athletes, is relatively young. At a conference in 2021, Trent Stellingwerff, senior adviser for innovation and research at the Canadian Sport Institute Pacific, mentioned that he conducted an informal audit of the scientific literature on nutrition to determine how well women were represented. The 2010 IOC Consensus Conference on Nutrition in Sport included discussion of five sex-based themes specifically related to sex-based differences in energetic intake, carbohydrate loading, eating disorder prevalence, body composition, and the female athlete triad. Nine years later, the number of sex-based themes discussed in the World Athletics (then the IAAF) Consensus Statement increased to twelve. It's an improvement, but there's still more work to be done.

In the 1960s and 1970s, researchers began investigating the relationship between diet, exercise, and women's health as more women took up physical activity and sports. There were anecdotal reports popping up that exercising women experienced abnormal menstrual cycles, which raised some red flags. It harkened back to concerns, from the turn of the twentieth century and earlier, that women's bodies were fragile and not suited to vigorous activity. All fingers pointed to exercise as the main culprit for the disrupted cycles.

It wasn't until the 1980s and 1990s that researchers started to connect the dots between menstrual health and nutrition. Studies led by exercise physiologist Barbara Drinkwater laid the foundation for what would become known as "the female athlete triad," which was initially described as three distinct conditions: amenorrhea (the absence of menstruation), poor bone health, and disordered eating. As researchers continued to untangle the triad's biological pathways, they noticed something interesting. In some cases, blood tests showed that women with menstrual cycle dysfunction also had abnormal levels of other biomarkers. Just looking at the lab results, you'd think they were starving.

The observation led Anne Loucks, then a professor at Ohio University, to wonder if nutrition played a more prominent role in the complex condition. If researchers manipulated dietary intake, would that lead to a change in reproductive function? To tease apart the independent effects of exercise and diet on hormones, Loucks led a series of seminal studies with regularly menstruating women who weren't athletes. Her hypothesis centered on the idea of energy availability, the pool of energy that's left over when you take the total calories consumed and subtract the calories burned through exercise. Like a tank of gas, it's the fuel on hand that the body draws on for functions like cellular maintenance, growth, thermoregulation, and reproduction.

In her experiments, Loucks altered energy availability by either restricting caloric intake or increasing the number of calories burned through exercise. When women ate a diet that met their daily energy needs but increased the amount they exercised, Loucks noticed changes to luteinizing hormone (LH), a hormone necessary for normal ovarian function. Regular surges of LH help keep the menstrual cycling ticking, but Loucks noticed in the experiments that the LH surges were suppressed, a sign of hormonal dysfunction. When women ate a reduced diet but didn't exercise, there was a similar dampening of LH pulses. What's more, Loucks and her team discovered that they could prevent the hormonal disruption if women recouped the calories they used for exercise. "Athletes could do all the exercise as long as they were willing to eat," Loucks says. Before these experiments, scientists weren't aware of this relationship between nutrition and reproductive health.

Loucks's findings were a significant turning point, demonstrating that exercise in and of itself wasn't the problem. Energy availability was the key determining factor. The discovery led to an evolution in the understanding of the female athlete triad, which is now described as three interrelated conditions involving energy deficiency, menstrual dysfunction, and impaired bone health. Each of these conditions can exist on a spectrum, from healthy to disordered. A person who menstruates doesn't need a clinical diagnosis of an eating disorder, amenorrhea, or osteoporosis to be worried. Even a slight disturbance in any one of these areas is cause for concern.

In recent years, a parallel track of research has emerged as doctors and scientists noticed other symptoms linked to insufficient nutrition beyond those described by the triad, including conditions related to cardiovascular health, immune health, gastrointestinal distress, and mood disorders. In 2014, the International Olympic Committee proposed the term "relative energy deficiency in sport" (RED-S) and outlined a model for this broader constellation of relationships between energy availability and health. While there's some contention between the triad and the RED-S camps, primarily around terminology and research methodology, both point to energy availability as the underlying culprit. Studies have found that nearly half of women athletes across more than forty different sports were at risk of low energy availability. And it's not just a problem for professional or collegiate athletes—experts say recreational and young athletes are also at risk.

The tricky thing is that there isn't a specific threshold where women enter dangerous territory; instead, it's a delicate balance that varies from person to person. There aren't always visible signs of a problem either. These shifts can happen under the surface and out of sight, even in people who maintain steady weight, body composition, and menstrual status. But the body intuitively knows when there's a deficit.

When former pro triathlete Jenna Parker moved to Australia to join a new coach and training group in 2009, she had her heart set on representing the United States in the Olympics. Knowing that Olympic athletes are held to a superhuman training standard, she recognized that she needed help cultivating her raw athleticism. While her increased training load and intensity paid off—Parker placed second at the U.S. Nationals and won the Pan American Cup that year—she thought about her weight a lot. It was the first time a coach regularly weighed her and tracked her body composition, pinching her skin between metal calipers to estimate her body fat percentage. Her coach used these numbers as a proxy for fitness. If her numbers didn't meet his expectations, he wouldn't let her race because, presumably, she wasn't fit enough. When Parker gained a few pounds during the off-season, he gave her an ultimatum: lose the weight or leave the group. That's when Parker became what she calls a "functional anorexic," meticulously chipping away at her weight by reining in how much she ate. "I couldn't control how fast I got better at triathlon," Parker told me. "But if I lost the weight and got my skin folds down, he couldn't kick me out of the group. It was the thing I could control."

By the end of the season, Parker's underfueling had taken its toll. The week before the 2010 Dextro Energy Triathlon in London, her entire body broke out in hives for no apparent reason. It was her body's last-ditch effort to warn her to stop pushing herself, but she raced anyway. When she crossed the finish line in fortieth position, she collapsed and began crying uncontrollably. It was her lowest-place finish in nearly four years, and she was broken—physically and mentally. Her doctor shut her down for the rest of the season.

The consequences of not eating enough are more than just a raging case of "hanger" or a bad race result. "I think there's a psychological component to it. When you start restricting calories, it's like your body can tell. It goes into fight-or-flight mode and doesn't want to lose the weight," Parker told me. When the body doesn't have enough energy to fuel itself, it quickly switches to conservation mode, diverting resources away from nonessential functions. It's akin to overdrawing your checking account, but with no overdraft protection. The hypothalamus and the pituitary gland begin to put the brakes on the production of hormones like estrogen and progesterone, leading to menstrual dysfunction. It also coordinates with the adrenal gland to pump out more of the stress hormone cortisol and with the thyroid to adjust levels of thyroid hormones. A whole host of other hormonal changes occur too, and the body responds by burning fewer calories, conserving fat stores, and sending signals to tell you that you're hungry and need to eat.

All of these changes have downstream effects. As the body scrambles to respond to its lower energy state, it ignites cascading reactions that can negatively affect both endurance and power athletes. Researchers have observed decreases in coordination, endurance performance, muscle strength, and training response along with increased injury risk, impaired judgment, irritability, and depression. In one study, energy deficient athletes with amenorrhea demonstrated slower reaction times, lower muscular strength, and decreased endurance compared to those with normal menstrual cycles and sufficient energy levels. Overall, women can be left feeling drained and tired all the time and stuck in a cycle of stale training and performance, which can lead to overtraining and burnout. For Parker, it took four months of rest and recovery after her 2010 breakdown before her body even entertained the thought of training again.

But doctors and researchers are most concerned about the long-term health implications of not eating enough. Women athletes with low energy availability experience four and a half times the risk of bone stress injuries and are at greater risk for early onset osteopenia and osteoporosis. When women underfuel, they don't experience the regular pulses of estrogen that are critical for bone building. This leaves them susceptible to low bone mineral density, altered bone architecture, and weaker bone tissue. Just five days of fewer calories can increase the rate of bone breakdown in women, even if their menstrual cycles appear normal. Lack of fuel can also trigger changes in the body that lead to higher levels of cholesterol and triglycerides in the blood, changes that can make an athlete's bloodwork resemble someone who is postmeno-pausal and at risk for heart disease. In some cases, blood vessels don't dilate readily or respond to the demands of the circulatory system, which can contribute to problems like high blood pressure and coronary artery disease.

Experts also warn against dramatic dips in energy during the day. Over the course of twenty-four hours, energy levels rise and fall depending on when you eat and exercise. Imagine your energy level shown on a graph where the y-axis shows energy levels and the x-axis represents time of day. Upon waking up, you're in a big hole. You're in negative territory on the y-axis (and below the x-axis) because the body has burned through a good chunk of your energy stores overnight. Once you eat breakfast, the curve starts to move up toward a more balanced state and even above the x-axis, signaling an energy surplus. When you exercise, the line dips down again as you begin to draw on your stored fuel to power your workout.

When researchers zoomed in and examined energy levels in one-hour increments, they found something interesting. Women endurance athletes who spent more time in an energy deficit throughout the day had suppressed metabolic rates and estrogen levels, higher cortisol levels, and menstrual dysfunction, even if they met their total nutritional needs over a twenty-four-hour period. What that means is that shorter-term periods of mismatched energy, even coming up as little as 300 calories short, matter. "Looking at the way that food, nutrition, and calories are partitioned across the day is absolutely part of the puzzle," Trent Stellingwerff says. While skimping on a few hundred calories a day doesn't sound like much—the equivalent of a medium apple and two tablespoons of peanut butter—it adds up. Over the course of a year, Stellingwerff says, it's comparable to not eating for an entire month.

While the studies of energy availability have largely centered on women, discussions in the field have opened doors to understanding the relationship between nutrition and hormonal disruptions across the spectrum of sex and gender, and scientists have noticed that men struggle too. Men can also experience poor bone health, as well as changes in reproductive hormone function as a result of not eating enough. In this case, they may have low testosterone and low sperm count. It's part of the reason behind the development of the broader RED-S framework and an updated model for the athlete triad, which now includes men in the diagnostic and clinical criteria and is called the Female and Male Athlete Triad.

The field of sports nutrition is confusing to navigate. There's a tremendous amount of information to wade through regarding what to eat (including how much and in what proportions); how to adjust those recommendations based on environmental conditions like high altitude and heat; and if and how supplements should be used. The current focus in the field seems to be caught up in the flourishes of how we eat and the quest to exploit any margin of gain. We focus on diets that promise to optimize performance and unlock a happier, healthier you. But along the way, we've lost sight of the fundamentals—basic nutrition and adequate energy. We've forgotten the importance of ensuring that we're simply eating enough food, and right now, many people are falling short. And without sufficient energy, the body can't function to the best of its ability.

Take intermittent fasting, which involves eating only during a designated window of time during the day. Depending on the protocol, this could mean fasting for anywhere from eight to twenty hours, or fasting every other day. Its popularity stems from promising research that shows time-restricted eating may help improve certain markers associated with chronic disease like cancer and diabetes, enhance weight loss, and improve cellular repair.

As this pattern of eating has gained traction, it has jumped over to the world of health and fitness. It's promoted on podcasts, in ads in running magazines, and on social media as a way to gain a competitive edge by spurring advantageous metabolic adaptations, particularly for endurance athletes. The idea is that athletes burn through carbohydrate stores pretty quickly during long-distance events—once these stores run out, the body has to turn to another source of fuel, namely fat. But humans are less efficient at metabolizing fat compared to carbohydrates. Theoretically, since intermittent fasting imposes conditions similar to a long run or ride, restricting eating to specific time periods can train the body to utilize fat more efficiently. When it comes to women, the thinking goes one step further. Since women tend to carry more body fat than men, they potentially have more stored fuel to use.

However, most studies that demonstrated health benefits were conducted in animals or with human participants who weren't active, many of whom were overweight or obese. When researchers took a closer look at the impact of intermittent fasting on high-intensity, endurance, and resistance exercise performance, it was a toss-up; approximately half of the studies showed no significant difference between exercising in a fasted versus fed state. What's more, women responded differently than men. In studies, men performed better when they fasted before endurance training while the *opposite* was true for women. They performed better when they ate before exercise.

Intermittent fasting is built around the premise of restricting calories and forcing the body to adapt to a state of stress. It's part of the reason people may lose weight and become better fat burners. While men athletes can get away with periods of not eating, women are more sensitive to a downturn in nutrition, according to Stacy Sims, exercise physiologist and author of *Roar*. Sims says the reason primarily has to do with reproduction. The body is smart—becoming pregnant and carrying a baby to term isn't ideal if there isn't enough food around. When energy availability is low, the body's hormone levels adjust and ultimately dampen the menstrual cycle. Add in a workout session (a stressor in and of itself) and it can lead to overtraining, fatigue, and injury.

"If you're not going to put more energy in above what the body needs to rebuild the tissues broken down during exercise, then it's a big hole," Sims says. After a fasted cardio session, women showed a higher level of stress and inflammation as well as signs of muscle tissue damage and breakdown, all of which can be counterproductive to an active life over the long-term. Even skipping breakfast can create enough of an energy deficit to harm health and performance.

Letting a clock determine when to eat or not eat doesn't account for the normal daily changes in energy expenditure and the hormonal shifts that naturally occur in women. We don't have the same needs day-to-day. "Why would we only need to eat within that window of time, no matter the day, when there's so much else going on in our bodies?" says Heather Caplan, a registered dietitian.

Similarly, the ketogenic, or keto, diet has become an increasingly popular way to attempt to boost performance. While low-carb diets have been around for many years, keto is *super* low-carb. It essentially turns the food pyramid on its head: eat mostly fat, a smidgen of protein, and barely any carbs (less than 50 grams a day). Like fasting, keto forces the body to switch from using sugar and starch as fuel to using fat. Some of that fat is transformed into ketones, which can cross the blood-brain barrier and supply the brain with energy in the absence of glucose.

The keto diet was originally developed to treat severe cases of epilepsy in children, but in the early 2010s it was picked up by lifestyle gurus and biohackers like Tim Ferriss, Dave Asprey, and Joe Rogan. Proponents raved about the diet's ability to sharpen the body's metabolic engine, improve body composition, and increase mental focus. The anecdotal reports created a cultlike following. As Michael Easter writes in *Men's Health*, "Keto thrives in the vortex of social media. It's highly viral because it's photogenic, offering swift results and dramatic befores and afters on Instagram."

When researchers examined the science behind the claims, the results weren't exactly a resounding endorsement, particularly for athletes. One of the most rigorous studies involved world-class race walkers, both women and men. They were divided into two groups during a three-and-a-half week training camp. One group was fed a low-carb, high-fat keto diet while the other group was fed a high-carb diet. Both diets were matched for energy, so energy availability wasn't an issue. After the training camp, all athletes resumed a normal diet with readily available carbohydrates for two and a half weeks before a 20-kilometer race. While the low-carb group showed signs that they had adapted to use more fat for fuel, they also used more oxygen, meaning they were less efficient. Their performance was 2.3 percent slower, while the high-carb group improved their performance by almost 5 percent.

Compared to men, women's bodies are more adept at flipping the switch from carb burner to fat burner. They already utilize a greater percentage of fat as fuel and there isn't much room for improvement. Women may also need more carbohydrates to keep the endocrine system humming. The hypothalamus is particularly sensitive to a lack of carbohydrates—even when overall energy is sufficient—which can set off a series of actions leading to increased signs of bone breakdown, hormonal dysfunction, and other indicators of RED-S. When Jennifer Goodall, a recreational athlete, and her husband decided to try the keto diet, they had markedly different experiences. Her husband was clear-headed and full of energy, while Goodall was exhausted. Her head was foggy, she lost muscle mass, and she never hit any personal bests. She couldn't ride her mountain bike for squat. Her heart rate was jacked, and her legs felt like crap. She kept thinking that if she just pushed harder and controlled her diet even more, her body would eventually adapt. But the stricter she was with her diet, the more her performance suffered.

As with intermittent fasting, a low-carb diet causes a downturn in nutrition that shifts the body into survival mode. Sims says that with the keto diet, "women get fatter and slower. They get tired and more anxious and depressed because all these functions that require carbohydrates aren't getting the carbohydrates they need." Low-carb diets also bump up cortisol levels in all people, making it harder to synthesize protein, build muscle, and recover.

Even plant-based diets can get into tricky territory if people don't pay attention to their basic nutrition and energy requirements. On the whole, the emphasis on eating nutrient-dense vegetables, fruits, legumes, and complex carbohydrates offers many benefits, particularly for active people. Preliminary research suggests that an antioxidant-rich diet reduces inflammation and potentially protects cardiovascular health, providing ample carbohydrates for the body as well. But vegan and vegetarian diets tend to consist of high-volume, fiber-rich food that can fill you up fast, even before you've consumed an adequate number of calories. "If you're not really thinking about the way you're putting your meals together, you could end up in a low energy state," says Laura Moretti Reece, registered dietitian with the Female Athlete Program at Boston Children's Hospital. For those who are active and choose to be plant-based, Reece advises getting regular bloodwork and consulting with a registered dietitian and physician to make sure you're getting all the energy, macronutrients, and micronutrients the body needs.

The same is true for diets like the Mediterranean diet and DASH diet, which are rich in whole foods and lean protein and are recommended for their wide-ranging health benefits. It's still possible to miss calories even when following a "good" diet, which can leave the body without the energy it needs to thrive.

While there is so much more we need to understand about the relationship between nutrition and non-male bodies, what we do know for certain is that meeting the body's energy needs is paramount, particularly for girls and women. Without sufficient energy to support the body's essential functions, you're building your fitness and athletic training atop a flimsy foundation, one that can easily topple over. In a recent paper published in the journal *Sports Medicine*, researchers from the Female Athlete Program at Boston Children's Hospital recommend thinking about your "hierarchy of nutritional needs" instead.

The number one priority should be building a strong base of calories to avoid long periods of time where energy supply dips into the red. It's fundamental. When active people keep their energy tank consistently topped off, it drives long-term training adaptation, and training adaptation leads to better performance. When junior elite swimmers consumed enough food to support their bodies during training and had regular menstrual cycles, they swam faster. Over the course of a twelve-week season, they improved their velocity during a 400-meter time trial by roughly 8 percent. Meanwhile, swimmers who exhibited menstrual dysfunction and didn't eat enough saw their performance decline by 10 percent.

Even a few hundred extra calories can help women avoid some of the pitfalls of low energy availability. Between 2006 and 2014, researchers from Penn State University recruited women to participate in a randomized control trial. Participants, all of whom experienced menstrual dysfunction, were asked to consume approximately 300 extra calories per day while maintaining their normal exercise routine. Compared to the control group, the women who ate more were more likely to regain their period during the research period.

While many people think that they need to limit what they eat to achieve a lean, athletic physique, that's not always the case. Among sixty-two elite athletes, researchers found that middle-distance runners were not only some of the leanest athletes, but they were also the ones who came the closest to eating enough to balance their energy needs. Rhythmic gymnasts, on the other hand, regularly dipped 600 calories below what their bodies needed in a day, and they had the highest percentage of body fat. It seems counterintuitive, but tipping the scales toward too little energy can rejigger metabolism and spur the body to conserve more fat. If the body is fed, it can relax.

It's a lesson Jenna Parker learned toward the end of her triathlon career. She started working with a new coach who believed that if she ate well and followed his training program, her body would naturally find its race weight (the weight and composition at which it performs best). "When you tell your body you're going to feed it all the time, it's like, Cool. I can work with you and I'm going to keep burning those calories," Parker told me. "You get out of that state of fear." She says the irony is she ended up racing at roughly the same weight she was when she was a member of the Australian group and broke out in hives. But this time, she was much healthier.

Once energy needs are prioritized and taken care of, then you can think about the next level of optimal nutrition—how to allocate carbohydrates, protein, and fat in your diet along with vitamin and mineral intake, tailored to your health and medical history. Women need to pay particular attention to micronutrient deficiencies like iron, calcium, and vitamin D. As you gain more experience in your sport and your training goals become more performance oriented, you can think about the timing of meals before, during, and after working out, as well as how the specific demands of your physical activity—how much, how hard, how long—may influence energy needs. It's critical to eat regular meals, especially breakfast, and snacks around exercise sessions to avoid wild fluctuations and dips in energy availability during the day.

Laura Moretti Reece likes to refer her clients to the Athlete's Plate. It's a visual tool created by the United States Olympic Committee's Food and Nutrition Services and the University of Colorado Colorado Springs' Sport Nutrition Graduate Program to show how meals can vary depending on training volume and intensity. There are training plates for easy, moderate, and hard training days, and each plate illustrates a different breakdown of whole grains, lean proteins, vegetables and fruits, and fats. "Nothing ever comes off the plate, but we might play with the portions of things as your level of activity is changing," Reece says. She's quick to point out that this plan includes carbs. If you want to get into the nitty-gritty details, a registered sports dietitian can help tailor a nutrition plan to your specific needs, goals, and activity levels.

It's only after all these other priorities have been addressed that hormonal profiles and other age-related factors may come into play and one can potentially consider dietary tweaks to account for a woman's physiology during different phases of the menstrual cycle or from using hormonal contraceptives. However, jumping to adopt highly specialized advice, without a solid foundation in place, could do more harm than good. As the researchers from Boston Children's Hospital write, "Failure to proceed in a graduated fashion can overwhelm the athlete and hinder adherence the same way that too advanced a training plan can lead to injury."

Until we get fueling right, we'll continue to keep active and athletic women from reaching their potential. Until we disentangle weight and body size from health and athletic potential, we'll continue to ask girls and women to do the impossible: Be active, but not too athletic. Live up to male standards of athletic achievement, but without fueling yourself properly. Eat, but not too much. Strive for success, but ignore the risk of long-term health problems.

It doesn't have to be a hopeless situation. "You can be healthy and also be incredibly good at whatever sport you're doing," Jenna Parker says. "We need to look at the female body and the differences we have not as a negative." And these physiological differences can give women a distinct advantage if we recognize and appreciate them in their own right.

5.

The Long Game

· · ● ● ● ● · ·

S arah Thomas just wanted to know one thing: Where was the
beach?

Twenty-four hours earlier, at midnight on September 15,
2019, the thirty-seven-year-old from Colorado stood on the shore in
southeast England, near Dover. The moon was bright, and the dark
water lapped gently over the rocks. A bright yellow light hung off the
back of her blue swimsuit and a green light was strapped to the top of
her gold swim cap. Thomas stood with her hand raised, signaling to the
observers on her support boat, the *Anastasia*, that she was ready. Then
she stepped into the English Channel and began her journey through
the choppy 65-degree water toward France, stroke by stroke.

Now, she was headed back to England, almost halfway through her
attempt to become the first person to cross the Channel four times
without stopping. She knew she was tantalizingly close to the coast.
More than anything, she longed to sit in the surf along the shoreline for
her allotted ten minutes before swimming back to France.

Around midnight, Thomas reached a concrete seawall on the English shore and couldn't swim any farther. She lifted her head and looked

toward her friend Elaine Howley, who swam nearby. As Thomas's safety swimmer, Howley was like a mobile pit crew. According to rules set by the Channel Swimming Association, Howley couldn't physically touch Thomas during the swim, but she could pass along anything Thomas might need. She carried a drawstring backpack filled with supplies, like extra lanolin to prevent chafing, the diaper rash cream Desitin that Thomas used as makeshift sunblock, and a bottle of warm water. Her other task was to guide Thomas the last 100 meters or so to the turn-around point at the end of each 21-mile traverse. This time, she had a third job—to tell Thomas there would be no beach landing.

The problem was that the clear night and good weather were perfect conditions to attempt a Channel crossing, attracting other swimmers and support boats to the area. There was too much traffic in the water and the pilot of the *Anastasia* couldn't safely thread around them and get Thomas to the beach. Instead, she had to swim northwest a few hundred yards and make do with the seawall. As the words sunk in, Thomas broke down.

Thomas is no stranger to spending multiple days swimming. She currently holds the world record for the longest unassisted, current-neutral open water swim, having ticked off 104.6 miles in Lake Champlain over 67 hours and 16 minutes in 2017, the first person to complete a continuous century swim. But this time, Thomas wasn't sure she had it in her. "I said to myself, 'A two-way is pretty solid. No one is going to be mad at me for just doing a two-way,'" she told me.

Thomas felt lousy from the start of her swim and couldn't get into a groove. A jellyfish stung her in the face. Her stomach churned with the choppy waves. She threw up during the first six hours and again around the thirteen- or fourteen-hour mark. She continued to vomit as she treaded water at the seawall. Howley asked Thomas if she could just swim until sunrise. Then they would reassess the situation. As much as Thomas wanted to quit, she knew she hadn't come for a double

Channel crossing. She came for a four-way crossing. Reluctantly, she started her third leg and headed back toward Cap Gris-Nez in France.

For the next six hours, she continued to vomit—a lot. Everyone aboard the *Anastasia* was worried. They didn't see how Thomas could continue without any fuel in her system and considered pulling her from the water. Eventually, her crew found a fix. Onboard, they had anti-nausea medication, left over from Thomas's treatment for stage 2 triple negative breast cancer, which she'd completed just a year earlier. Fortunately, it worked. Once her stomach settled, Thomas began to recoup the calories she lost by nibbling on the crew's peanut M&M's. Every once in a while, she touched a spot near her hip where she carried a tiny pebble from her hometown reservoir in Colorado. The stone reminded her that she was okay, even in the gigantic expanse of the English Channel.

Finally, as Thomas made the turn to start her final leg, she felt calm, solely focused on the water and the rhythm of her arms and breath. After 54 hours and 10 minutes, Thomas was back where she started. She hauled herself up onto the beach and sat on the shore in disbelief, her body and face still covered in goopy white Desitin.

Crossing the English Channel is an iconic achievement. Along with the Catalina Channel swim in California and the swim around Manhattan in New York, it's one of the jewels in the Triple Crown of open water swimming. Crossing once is a victory. To swim it multiple times is a major triumph. Since 1875, 2,245 people have completed a one-way crossing and 35 have completed a two-way crossing. Only 4 swimmers have swum the Channel three times in a row (two men and two women). With her swim, Thomas became the first person to complete four crossings nonstop. In doing so, she followed in the footsteps of other pioneering women swimmers like Gertrude Ederle, who in 1926 at age nineteen became the first woman and sixth person to cross the Channel. And Ederle swam it nearly two hours faster than any of the five men before her, despite foul weather conditions.

Thomas never expected to become a marathon swimmer. Though she grew up swimming and walked on to the University of Connecticut swim team, she says she was a pretty average athlete. And back in the early 2000s, there weren't many opportunities to test her endurance beyond the mile event. But even then, there were signs that she should have been racing longer distances. In practice, she'd hit her stride just as the team began its cooldown, which drove her teammates crazy. They thought she was sandbagging, deliberately taking it easy during the first half of practice only to swim harder at the end to make the other swimmers look bad. In actuality, it just took Thomas that long to warm up.

It wasn't until after college that Thomas found open water swimming. At the suggestion of some friends, she signed up for the 2007 Horsetooth Open Water Swim, a 10K race outside Fort Collins, Colorado, and "just totally fell in love, like head over heels in love with open water," she told me. Without the confines of a pool, Thomas didn't have to worry about flip turns. She could just swim. She placed fifth out of fifty-six athletes. The next year, she won the event outright and thought to herself, "I might actually be kind of good at this." Since then, she's completed the Triple Crown and charted multiple firsts—the first back-and-forth swims of Lake Tahoe and Lake Memphremagog (located between Vermont and Canada), and the first swim of Lake Powell on the border of Arizona and Utah—in addition to her record-setting swim in Lake Champlain and her four-way English Channel crossing.

For Thomas, the longer the distance, the better she feels. Something clicks and she, in turn, shines. Thomas isn't the only woman making her mark in ultra-distance events. Also in 2019, Jasmin Paris crossed the finish line first at the 268-mile Montane Spine Race, often dubbed "the most brutal race in Britain," fifteen hours ahead of her next competitor, all while expressing breastmilk for her daughter Rowan at aid stations along the way. Fiona Kolbinger won the Transcontinental Race, becoming the first woman to win the ultracycling event, finishing nearly eight hours ahead of second-place finisher Ben Davies. Maggie Guterl was the last runner standing at Big Dog's Backyard Ultra in Bell Buckle, Tennessee, after having run 250 miles over sixty hours to become the first woman to win the event.

These staggering feats of endurance are a dramatic shift from the days when the prevailing wisdom was that a woman's uterus would fall out if she ran too much. The media has seized upon these stories. But implicit in the headlines, which range from "The Woman Who Outruns the Men, 200 Miles at a Time" in *The New York Times* to "Could Women Run Faster Than Men?" in *The Telegraph*, is the assumption that these athletic achievements are an anomaly, that women are somehow exceeding what's expected of their bodies, especially if they outperform men. The disdain runs so deep that there's even a widely used phrase for the phenomenon—getting "chicked."

But whenever I came across these articles or news segments, I couldn't help but wonder: Could this be an area where women are uniquely adept and have an advantage over men?

Humans have long kept track of who's the best—in the world, on this course, in this pool, in this incredibly niche event. Those benchmarks largely reflect what men have accomplished in sports. Men's achievements have become the de facto measuring stick and framework to organize and understand athletic performance and progression, and women are judged by this standard too. Women have never been given the space to test their potential and to set their own benchmarks without the weight of expectations that have been tainted by what men have accomplished or misconceptions about women's bodies.

What could women achieve if they were given a blank slate and nothing to compare themselves to? What if women were given the freedom to launch an entirely different athletic trajectory than men?

In the early 1990s, physiology professors Brian Whipp and Susan Ward looked at the progression of men's and women's world records in running events ranging from the 200 meters through the marathon. They found that men lowered their times at a fairly predictable rate across all events. Yet the rate of improvement for women, particularly in the marathon, was much steeper. Based on the data, they predicted that the gap in times between women and men in the marathon would cease to exist by 1998.

It didn't quite play out that way. In 1998, Ronaldo da Costa and Tegla Loroupe set new men's and women's marathon world records, but the gap between the record-breaking times was still more than fourteen minutes. And though the gap has edged slightly closer to twelve minutes between the current world records, there's still a 9.7 percent difference today.

One reason for Whipp and Ward's overconfidence was that they treated race results as purely a mathematical equation and assumed that velocity would increase at a consistent rate. They failed to account for women's late entry to long-distance running; women weren't allowed to compete in the marathon until the 1970s. It made sense that women's performances improved by leaps and bounds, particularly in the first few decades of participation, before leveling off.

Whipp and Ward also ignored fundamental differences in anatomy and physiology between men and women that could influence athletic performance. Before puberty, girls and boys are more or less athletic equals. But once sex hormones, particularly testosterone, flood the bodies of adolescent boys, everything starts to grow—hearts, lungs, muscles, and limbs. Their bodies lean out, which translates to more strength, power, and speed. Men also typically score higher on measures of aerobic capacity—how much oxygen they're able to take up and use during exercise. With bigger lungs, they can take in more oxygen. With bigger hearts and higher hemoglobin levels, they can pump a greater amount of oxygen-rich blood to their muscles. With bigger muscles, they can extract more oxygen from the blood.

While estrogen can influence factors related to training adaptation, performance, and strength, its influence is less potent. "There are always going to be these fundamental differences between males and females where the best male, under the right conditions, will outdo the best female," says Sandra Hunter, director of the Athletic and Human Performance Research Center at Marquette University. On average, across athletic disciplines, women's records are 9 to 12 percent lower than men's records, whether it's sprinting, jumping, throwing, or distance events.

However, that's not the end of the story. As distances increase, some of the anatomical and physiological advantages enjoyed by men begin to wash out. Particularly in ultra-distance events—those that exceed six hours or running events longer than 26.2 miles—outcomes are less dependent on physiology or cardiovascular capacity alone, says Nicholas Tiller, an exercise physiologist who studies how the body responds to extreme endurance exercise and an ultramarathon runner himself. As distances stretch out over 50, 100, or 200 miles, athletes must account for and manage many more variables. Weather, nutrition, gastrointestinal health, fatigue, pain, and psychology all start to carry more weight. "If you start suffering from gastrointestinal distress and you start feeling nauseated, it doesn't matter what your VO_2 max is. It doesn't matter how strong you are. It doesn't matter how quickly you can run if you can't even stand because you can't get the calories in," Tiller told me.

And it seems like women may be better able to juggle the multiple factors that go hand in hand with long distances. In 2020, researchers examined more than five million results from nearly 15,500 ultrarunning events to determine the average pace and finishing time across all participants. They found that as distance increased, the gap between women and men narrowed. While women were, on average, 11.1 percent slower than men in the marathon, that percentage dropped to 3.7 percent for 50-mile races and just 0.25 percent for 100-mile races. At distances over 195 miles, women were 0.6 percent *faster* than men. In other words, the average woman was faster than the average man in superlong races.

As I sifted through the studies and anecdotes, I wanted to know if this trend reflected a true sex difference or if it could be sampling bias. Are the impressive performances by women ultra-athletes just a result of the best of the best lining up on the start line, or are they representative of what the larger population of women is capable of?

Currently, it seems like it might be a case of sampling bias. Women constitute a small segment of entrants to ultra-events (they make up only 23 percent of total ultramarathon participants). Those who choose to race long distances are likely well-trained athletes, skewing the average woman's performance toward the top of the heap. On the men's side, the larger number of competitors is more likely to represent a wider range of experience and abilities, contributing to the men's lower overall average times. Still, at the 2021 Western States Endurance Run, the oldest and one of the most prestigious 100-mile trail races, brutal temperatures whittled the field down to its lowest finishing rate in more than ten years. Yet three women finished in the top ten overall—a first in race history—and fifteen of the top thirty finishers were women. These numbers are hard to ignore.

Frankly, despite the curiosity, asking if and when women will outperform men isn't the most salient question. Evaluating women's results against the men's standings continues to suggest that women are less than men and only worthy of accolades if they live up to standards set by and for men. It's like saying they performed "good for girls."

Instead, women's achievements should be acknowledged as excellent on their own. This is not to say that we need to lower the bar when it comes to women's sports. Rather, we need to shift the narrative to focus on women wholly rather than forcing them to measure up to a male paradigm. We need to recognize and celebrate women's unique abilities and lived experiences. Then the question becomes, What makes women well suited to run, bike, hike, and swim long distances? And why do they tend to excel when events get longer and longer?

If there's one person who fits the mold for a textbook endurance athlete, it might be ultrarunner Camille Herron. You can't miss her tall frame in races on the roads or the trails. She's usually at the front of the pack, her blond hair hanging loose around her shoulders. While her competitors may be grimacing, Herron is most likely smiling. She just looks like she's having fun—like there's nothing she'd rather do than log mile after mile after mile. Herron says she was born to run and her talent for long footraces has resulted in multiple American and world records, from 50-mile races to events that spanned twenty-four hours, world championship titles, and outright wins against all competitors. She even holds the Guinness World Record for the fastest marathon run by a woman in a superhero costume (she dressed as Spider-Man).

Running wasn't Herron's first love while growing up in Oklahoma. She wanted to follow in the footsteps of her father and grandfather, both of whom had played basketball for Oklahoma State University. But in middle school, Herron thought she was too short to play ball. She joined the cross-country and track teams instead. While she claims she was the worst sprinter, hurdler, and jumper, she could run for a *long* time. In high school, she made All-State three times in cross-country and won state titles on the track for the 1600 meters, 3200 meters, and the 4×800-meter relay.

While Herron had natural endurance and tenacity that suited the sport of running, as a young athlete, she didn't feel like her training matched her physiology. She thought her high school coaches, whom she describes as football guys who happened to oversee the running teams, were trying to work on her weaknesses. They knew she had a knack for longer runs but she needed to get faster. She remembers always running speed workouts but never learning how to run easy or recover.

Herron may have been on to something. Despite her success, her body was prone to injury. After shooting up nine inches, she developed Osgood-Schlatter disease, a condition that causes knee inflammation, particularly in growing adolescents. She had seven stress fractures during her high school and short collegiate career at the University of Tulsa. Her injuries forced her to take a medical hardship waiver in college and quit running competitively. "I really just think my muscles needed to catch up with my bone growth," she says.

Herron's body wasn't built for the high-intensity work her high school coaches prescribed. Instead, she thrived on longer runs. The turning point came in her later years of college, thanks to a poster signed by two-time Olympic marathon medalist Frank Shorter that said "Run for stress release." It shifted Herron's mindset. She began to treat running as a way to decompress and relax, not just a means to grind out miles. She did what felt good, which meant running slowly. And a lot. Without realizing it, by the time she was in her early twenties, she was logging 70-mile weeks. "By that point, I had grown into my body, so my muscles had caught up to my bones," she says. Her then boyfriend (now husband) Conor Holt, an All-American runner for the University of Oklahoma, began coaching Herron, and she thrived. She spent ten years training for the marathon, qualifying for three Olympic marathon trials and representing the United States at the Pan American Games.

While most people are pretty wiped after covering 26.2 miles, Herron says the distance felt short to her, "like a sprint." It was a hint that maybe she had more in her. Everything clicked when she completed the

2015 MadCity 100K. At age thirty-three, it was her first crack at the distance. "I felt like Billy Elliot doing ballet for the first time," she told me. The longer she ran, the more it felt like her body and mind woke up. "It's like I put on a jet pack at forty miles," she says, a point in the race when most people are fatigued. She started catching the guys too, some of whom had notched faster marathon times than she had. She took the women's title, which also doubled as the USA Track & Field 100K National Championship race.

Herron then won both the International Association of Ultrarunners (IAU) Women's 50K and 100K World Championships, becoming the first runner to win both titles in the same year. In 2017, she was the third American to win the storied Comrades Marathon in South Africa. Later that year, in her 100-mile debut at the Tunnel Hill 100, she set a world record, crushing the previous record by over an hour while winning the event outright. She followed up that performance by breaking the twenty-four-hour world record and she broke her own 100-mile world record in February 2022. She hasn't stopped since stringing together her first 100-mile week in 2006, averaging triple-digit mileage every week for seventeen years. A few months after her fortieth birthday, she clocked 100,000 lifetime miles.

What makes Herron so good at ultrarunning is largely her physiology and, more specifically, her aerobic capacity. Her body is primed for high-volume training. In grad school, her VO_2 max was measured in the high 60s, a level that indicates she has a pretty big aerobic engine. It's a proxy for the maximum amount of oxygen you're able to take up and pump to your muscles and is one of three factors that does a pretty good job of predicting performance in endurance sports.

While Herron hasn't had her muscles biopsied, she says she believes she's made up of "probably like ninety-nine percent slow-twitch" muscle fibers—the ones that power athletes through long, steady-state exercise and are less prone to muscular fatigue. Unsurprisingly, the ability to withstand fatigue can be a pivotal factor in whether an athlete crosses the finish line. And women, on average, tend to have a greater distribution of slow-twitch, or type I, muscle fibers. On the whole, these fibers are suffused with capillaries, the tiny blood vessels that carry oxygenated blood. They have more mitochondria too, the cell's energy powerhouse. That means slow-twitch fibers have everything they need to keep going and going . . . and going. Men, on the other hand, have a greater proportion of fast-twitch or type II fibers, which are better suited for speed and powerful bursts of movement, in part explaining men's historic dominance in traditional sports. But with a lower concentration of capillaries and mitochondria, these muscle fibers also tucker out faster.

Sandra Hunter has spent the better part of the last twenty years studying muscle function in women and creating models to explain why muscles are susceptible to fatigue. In the lab, she has observed that women tire less quickly during specific kinds of exercise. For example, she asked women and men volunteers to perform intermittent, static muscle contractions—repeatedly engaging their arm muscles at 50 percent of their maximum strength for six seconds, then resting for four seconds, until they were exhausted. The men quit the steady-state activity after an average of 8.5 minutes while the women lasted approximately 23.5 minutes—nearly three times longer.

Hunter says the discrepancies aren't due to a person's motivation or ability to activate their muscles. Instead, she says it comes down to the distinct composition of muscle fibers, and women have fibers that don't fatigue as fast. But Hunter says the greater muscle mass and muscle fiber diameter found in men might play a role too. When these stronger muscles are engaged, they might exert more pressure inside the muscle. The rise in internal pressure could restrict or cut off blood flow and oxygen, which would cause the muscle to tire quickly. Other studies in the field have found that women accumulated less waste product and blood lactate in their muscles, which may also contribute to their ability to withstand fatigue.

While interesting, these findings don't really tell us much about what might happen in the real world. In 2012, a team of Canadian and French researchers set out to determine if women were still more resistant to fatigue after an actual ultra-endurance exercise session. They recruited ten men and ten women who were participating in the Ultra-Trail du Mont-Blanc (UTMB) in the Alps, the Super Bowl of the ultra-running world. After the 110-kilometer race (the normally 171-kilometer course was shortened due to wintery conditions), the muscles in the men's calves and thighs were more exhausted compared to the women. The finding confirmed what Hunter saw in her lab studies: Women tire less than men after long, steady-state physical activity.

Aerobic capacity and muscle endurance are only part of the physiology puzzle. Another key factor in an athlete's success is how well their body uses carbohydrates and fat for fuel. While carbohydrates are a readily available source of energy for most activities, the body simply can't store enough carbs to keep an athlete going for extended periods of time. At some point, it must turn to fat to power working muscles, and the ability to burn fat efficiently matters in ultra-endurance sports.

Herron has a preternatural awareness of her body's energy and hydration needs, rarely dipping into a depleted state where her body is using more energy than she's taking in. While most people weigh less after a race because of the number of calories they expend and water weight lost through sweat, Herron says she often weighs the same, even after a twenty-four-hour race. She's never skipped her period either, a critical marker of hormonal and bone health linked to adequate nutrition. Herron thinks it's partly because her body is "probably wired for fat metabolism."

If an athlete is more adept at accessing and burning fat, their body may be better primed to sustain longer bouts of aerobic exercise because, theoretically, they won't run out of gas as quickly. Fat, with its greater energy density, offers a bigger bang for the buck compared to carbohydrates. Evidence suggests that women are more metabolically nimble, capable of switching from burning carbohydrates to burning fat, especially during the second half of the menstrual cycle. It turns out that women have more of a specific enzyme needed to facilitate the breakdown of long and medium fatty acid chains. Compared to men, women break down and metabolize fat at a rate that's up to 56 percent greater over a twenty-four-hour period, regardless of physical activity level. Even at submaximal efforts typical of endurance events, some research suggests that men use as much as 25 percent more of their stored carbohydrates.

Nicholas Tiller says this ability to utilize fat efficiently has "knock-on implications." To keep moving forward, ultra-athletes must keep their bodies regularly topped off with calories. Overall, woman ultra-athletes don't need to take in as many calories compared to men because of their smaller body mass. By not ingesting as much food, gels, bars, and sports drinks, women potentially dodge a bullet when it comes to meddlesome gastrointestinal issues like cramping, vomiting, nausea, loose stools, and bloating, which are frequently cited as one of the top two reasons athletes drop out of a race or don't perform as well as expected. Plus, if an athlete utilizes fat for energy, like women seem to do, they can reduce the risk of glycogen depletion—when they've used up their critical stores of carbohydrates and hit the dreaded "wall."

Certain characteristics of women's physiology may also provide an advantage in marathon swimming. Women tend to have a higher body fat composition in comparison to men, something that's considered a disadvantage in most other sports. But in swimming, higher body fat percentage translates to added buoyancy, meaning athletes don't have to work as hard to stay afloat.

The distribution of fat across the body matters too. "We know that males tend to hold more of their body fat around their midsection as visceral fat," says Tiller, whereas women tend to carry body fat around the lower body and hips. It's believed that women's on-average shorter limbs and stature makes it easier to assume a streamlined position. While it may not seem like a big deal, these slight differences may contribute to lower drag, allowing a swimmer to glide through the water with less resistance, requiring less energy to power their efforts. That advantage is evident in the record books. Over an eighty-seven-year period, the fastest woman to ever cross the Catalina Channel in California bested the fastest man by twenty-two minutes. When it comes to swimming a lap around the island of Manhattan, the top ten women were 12 to 14 percent quicker than the top ten men based on thirty years of finishing times.

Both Hunter and Tiller caution that when it comes to sex-based differences in endurance sports, more research is needed. It's an area of study that scientists continue to tease apart. While some studies show a definitive effect, others report no significant findings. In part it's because carrying out research on endurance sports, particularly with women athletes, is hard. For studies conducted in laboratories, scientists often default to men as participants for all the reasons discussed in earlier chapters, including the complications (and costs) related to controlling for the menstrual cycle. With field studies, scientists have a hard time getting people to participate in general. "They've been training all year for one race. They don't want to potentially compromise or sacrifice their race performance to help out someone who's doing some research," Tiller says. Lower participation rates of women in races make it even harder to recruit these athletes. With fewer (or no) women taking part in studies, scientists can't examine direct sex differences.

Since ultra-races often take place on high-alpine trails, in desert landscapes, and on other difficult terrain, scientists must consider the event's location and logistics—and how that might affect data collection.

If races occur in remote environments, it could be anywhere from an hour to a day or more for athletes to travel from the finish line to a lab. "At that point, there's no point taking any measurements because any of the acute, short-term effects are going to have worn off. You need to take these measurements as close to the finish line as possible," Tiller says.

In his studies, Tiller has done everything from camping out in the field for ten days to conducting tests with athletes in the back of a bus while traveling from the finish line back to base camp. He laughs as he describes what's supposed to be high-level scientific research. It also influences the type of measurements scientists can collect in the field. It's relatively easy to measure blood pressure and heart rate, take blood samples, and perform breathing tests. However, he says, these are superficial tests and "none of this is giving us a lot of information about how the body is responding" like CT scans might provide.

While there are some compelling reasons for women's success in ultra-endurance activities, researchers are still on the hunt for the X factor. As physiological advantages begin to fade away with longer distances, athletes say the psychology of sports and mental stamina begins to take on more importance. Maybe it's all in the mind?

Courtney Dauwalter has seen some pretty wacky things out on the trails. After running many hours through the mountains, her mind starts to play tricks on her. Puppets on a swing set. A pterodactyl and a giraffe. A woman churning butter. (Dauwalter waved at her.) While most people might freak out at hallucinations or the first sign that they're hitting the pain cave—the bleak point in a race or workout when everything hurts and feels impossibly hard—Dauwalter greets them like old friends. "Instead of trying to avoid it or prolong it or stay away from it as long as possible, it's like, Let's get to the entrance of that cave. Let's go in," she told me. For her, the real work of an ultramarathon isn't in the first 50 or 100 miles of a race. The real work comes when she desperately wants to stop.

Dauwalter's approach to the pain cave might seem counterintuitive to some. To others, it's indicative of the cult of suffering that's often endemic to endurance sports, where athletes voluntarily expose themselves to tremendous discomfort. But Dauwalter's ability to keep moving in the face of screaming, aching muscles is less about mental toughness per se or a masochistic desire to hurt. Instead, her interest in the pain cave is rooted in curiosity—about herself, her physical ability, the process of doing something audacious, and how she handles herself in those moments. It's something she learned as a Nordic ski racer growing up near Minneapolis. "Our coach always taught us that there's more gears than we think there are and that we can just keep cranking it up a notch even when we're hurting," she says. So, when she got into ultrarunning and arrived at the entrance of the pain cave, it felt familiar. She'd been here before and knew that if she entered the cave, she'd be okay.

That curiosity may be what sets her—and other women endurance athletes—apart. It's what drives Dauwalter's pursuit of running really long distances, not the podiums or course records (although she has plenty of both). Dauwalter is one of ultrarunning's most dominant athletes, having won some of the sport's most prestigious races, including UTMB and Western States. "I just keep realizing how little we know, and we haven't found a limit yet. How cool is it that I get to pursue that and try to find out a little bit more?" she says.

Dauwalter's fascination with the mental game started after her first attempt at 100 miles. It didn't go as planned, and she dropped out at mile 60 of Colorado's Run Rabbit Run 100 in 2012. She was exhausted. The pain in her legs was unlike anything she'd ever experienced before. It was "alarming," she told me. From there, she spiraled, thinking, "You can't do this. What a joke. You thought you were fit enough to run one hundred miles?" It was like her brain kept catching on a scratch on a vinyl record, repeating the words over and over, and she wasn't able to pick up the needle to move it to another track.

Surrounded by darkness, she convinced herself she wasn't cut out for ultramarathons and dropped out when she arrived at the next aid station. While waiting to be picked up, she watched other athletes rest and refuel. "They looked physically destroyed, but they still went about their business and then they kept moving forward," she recalls. Dauwalter doesn't say she regrets not finishing the race, but the experience triggered a fierce determination in her. Next time, she wasn't going to quit. She was going to be a 100-mile runner. A year later, she stood at the start of the Superior Fall Trail Race 100 in northern Minnesota. She finished in seventh place overall and took second place in the women's race.

After her first few ultras, Dauwalter realized how many puzzle pieces are in play on race day (or days). "You don't get to just knock it out of the park every time. There's a lot of finagling when you're out there for thirty hours or whatever," she says. Each training cycle and race is like a mini experiment where she figures out what went right and wrong. While she says she doesn't have "a whiteboard with everything I've figured out," she tucks away that information into one of the many filing cabinets in her brain. The next time she preps for a race or runs into trouble during an event, she can draw upon strategies that worked in the past.

And she thinks anyone can compete. "You just have to use your pieces better," she says. While a man may have more muscle mass or greater lung capacity, maybe Dauwalter can be more patient. Maybe she's more stubborn. Maybe she digs deeper into her pain cave. Or maybe she just uses the tools in her arsenal in a unique way. This is not to discount Dauwalter's training or physical ability—she's an incredible athlete. But it's the ability to juggle and deploy multiple puzzle pieces a little bit differently that may explain why women compete well in long races.

There's no doubt that psychology is a critical part of sports. Studies have shown that strategies like imagery, self-talk, and goal setting help improve performance, while mental fatigue can diminish it. But the mind may play an even bigger role in setting the limits of endurance specifically. Exercise physiologist Samuele Marcora studies the connection between the brain and endurance performance. He's developed what's called "the psycho-biological model of fatigue." Marcora believes there's a complex interaction between what is physically happening in the body and factors like motivation and mental fatigue. Together, they sway perception of effort. Ultimately, the choice to continue to move forward or stop is a conscious one and depends on whether the effort feels hard or easy. When a run or ride feels easy, or you can override the feeling of pain for a little while, you keep going.

While most ultra-endurance athletes share two key personality traits—mental toughness and a belief in one's ability to succeed—not much is known about women ultra-athletes in particular. To begin to bridge the gap, researchers from the College of Western Idaho and Boise State University surveyed 344 women ultrarunners, mostly from the United States. Many of those surveyed worked full-time and trained over twelve hours a week while managing family, relationships, and childcare responsibilities. (It's interesting to note that the researchers classified full-time students and stay-at-home mothers as full-time workers, since their hours are often the equivalent of full-time employment or more.) They found that women tend to be driven by health-related reasons. Unlike men, they are less focused on the competitive aspect of the sport and are spurred on by personal growth rather than recognition. These findings mirror other research conducted with women mountain bikers, who similarly valued the personal challenge and feeling of self-esteem that go along with sports.

"Intrinsic motivation and understanding your reason for doing" an event is important, says Carla Meijen, a sports psychologist whose research examines the psychological aspects of endurance performance. It's easier to quit when things get hard if your mind is set on a specific outcome just for the accolades. What's the point of continuing if you're not going to hit your goal time? "If you do have a clear reason why you're doing something, it gives you an extra boost or motivation to keep going," Meijen told me. Some also say that women just cope with physical distress better than men, thanks to experiences like childbirth and menstrual cycle pain. While there's anecdotal evidence to support this idea, there's no hard science to back it up. However, those experiences—and having survived them—may shift a woman's perspective on physical discomfort.

The desire to test one's potential compels other endurance athletes like Sarah Thomas. Thomas doesn't swim long distances for fame or to crush her competition. For her, these endeavors are designed to answer one question: Did she give it her all? She acknowledged she's not the fastest Channel swimmer, but she believed she could swim a four-way crossing. She also just really likes to swim—being alone in the water and knowing she powered herself incredible distances with just her arms and legs. That's not to say Thomas didn't want to quit during dark times in the water (like when she vomited for six hours), but her mindset and focus on the process may be a key reason she was able to weather those moments.

And it's not like Thomas jumped from swimming a 10K to traversing the English Channel four times. "It would never even occur to me to try and accomplish something like that if I hadn't done the years of work beforehand," she says. From one swim to the next, Thomas methodically raised the bar, building her knowledge and allowing her body to acclimate to longer distances. Every hurdle bolstered her confidence. When it was time to tackle a new distance, it didn't even occur to her that she could fail.

It's this difference in ego that Thomas has witnessed between women and men. "If I'm going to sacrifice time with my family and my career, I want to be really prepared to make sure that I'm successful, whereas I think guys, generally, have a tendency to wing it a little bit more," she says. Not only that, Thomas has multiple contingency plans to get her through the rough spots. Like Dauwalter's filing cabinets in her brain, having those "files" in place ultimately reduces the energy needed to problem-solve in the moment so Thomas can focus on moving forward.

Thomas's hunch may have some truth to it. Researchers from fields such as finance and education report that men tend to be more over-confident compared to their women counterparts, a tendency that carries over to sports too. When runners were asked to predict their finish times prior to the 2013 Houston Marathon, men consistently overestimated their abilities by an average of 9 percent, whereas women overshot their predictions by 7.6 percent. In the race, men held a fast pace early on, reflecting their ambitious mindset, but slowed down in the latter stages. Women, on the other hand, kept a steady pace throughout, which is associated with better race-day performance. Other studies have also found that women are better at holding a consistent pace in a marathon compared to men. These trends hold true even at ultra-distances.

It can be hard to determine if the differences in pacing are due to psychological factors, like a propensity for risk-taking behavior, or physiological considerations, like glycogen depletion or muscular fatigue. To try to disentangle these factors, researchers looked at pacing and results for the 2008 to 2013 BOLDERBoulder event, a 10-kilometer road race in Boulder, Colorado. Unlike a marathon, the 10K race is 6.2 miles, a distance where athletes are not likely to run out of fuel or become super fatigued before reaching the finish line, thus limiting the potential influence of physiology on pacing. They found that the odds of the fastest women maintaining an even pace were between 1.36 and 1.96 times greater compared to men (depending on how researchers measured pacing), indicating that decision-making—not physiology—likely plays a role. While the jury's still out on why women and men perform the way they do, it's clear there's a psychological component that comes into play.

And success begets success. "I've seen a shift in other women, and I feel like other women are getting that confidence to step it up and to not hold back," Camille Herron says. "More women feel like they can actually compete with the men and they're not afraid of that."

When I come back to the question of whether or not women are uniquely suited to traverse long distances, part of me wants there to be a simple answer, that the key to endurance performance can be boiled down to specific aspects of female physiology or women's mental toughness and high pain tolerance. Plain and simple. Then it would be easier to pinpoint the ways we can optimize performance for those who want to do better and create a formula to get there.

But it's clear that endurance is far from straightforward. There is a complex array of factors, and at times it can feel like you're shuffling a card deck, hoping you're dealt a winning hand on race day. But what research and athletes themselves continue to discover is that a winning hand doesn't look the same for women and men, especially for extreme distance events. Those winning hands may include a wider, more varied combination of cards. In the case of women, they may even include cards from the discard pile that no one thought to consider before.

When we start to expand the model of athletic performance beyond the traditional male framework, we add more cards to the deck, reframing what is possible for all people.

In order to increase our understanding of what's going on—physiologically and psychologically—and help athletes and coaches think more strategically, we need to learn more, in the lab and the field. We need studies that will help us interpret the relationships between the myriad variables at play across the gender spectrum, as well as between cisgender women. We also need to continue to support women in ultra-sports by rethinking policies like race-qualification requirements and the lack of pregnancy deferrals, which can discourage women from participating, and creating more open and supportive environments for diverse athletes. As women's participation numbers continue to edge up, along with the participation of nonbinary and transgender athletes, we'll keep learning more about the capacities and boundaries of what all people can accomplish in endurance sports.

Ultimately, reimagining the deck of cards we're playing with and playing with a bigger deck help debunk long-held myths about women's bodies and women's place in sports. They allow us to think more thoughtfully about what women are truly capable of. And endurance isn't the only area of sports where such a perspective shift can help bolster active women's health and performance.

6.

The Dreaded Female Body

• • ● ● ● ● • •

Briana Scurry doesn't recall the hit that ended her storied soccer career. On April 25, 2010, the World Cup champion was tending goal for the Washington Freedom, part of the former Women's Professional Soccer league, in an away game against the Philadelphia Independence. With about ten minutes left in the first half, Independence midfielder Lori Lindsey ripped a low, hard ball toward goal. Scurry crouched down and moved to her left to block the shot. Then she was on the ground bundled over the ball with Independence forward Lianne Sanderson on top of her. She had no idea how she'd stopped the ball from crossing the goal line or how she and Sanderson had collided.

What Scurry doesn't remember (and didn't see) was Sanderson's approach from her right. The forward tried to sneak in and nip the ball away before Scurry could secure it. But Sanderson was a fraction of a second too late. The two knocked into each other, and Sanderson's knee hit Scurry's head near her right temple. When the goalkeeper stood up, the names on the back of the jerseys were fuzzy, and the field seemed to tilt at weird angles. She played out the rest of the half, astonishingly blocking a couple more shots on goal. Scurry told me she just "reached up and tried to grab it somewhere in the front" as the black-and-white orb traveled toward her in what looked like stop-motion animation. When the referee blew the whistle for halftime, Scurry walked to the sideline, leaning to her left. The trainer met her, took her hand, and asked if she was okay. Scurry answered, "No," and walked off the pitch. She didn't know then that she had just played the last game of her career.

Scurry adores soccer and isn't blind to its risks, especially as a goalkeeper who is trying to stop players as they barrel toward her. "It's an absolute catastrophe waiting to happen," she laughs. Still, she wouldn't change her choice to play, despite the lunacy inherent in her position. Scurry had been hit plenty of times before and recovered with no lingering problems. "I had Abby Wambach fall on my head once during one-v.-ones with the national team in training camp," she says. Scurry face-planted into the grass and missed a day or two of training.

The thing that made Scurry one of the best goalkeepers in the game was her mind. She recognized patterns in the blink of an eye, figuring out how to block an opponent's approach on goal. Her kinesthetic sense—her ability to know where her body is in space—was off the charts. It's a sixth sense most athletes are familiar with and one that Scurry has had since she was a little girl.

Years earlier, on a sweltering day in July 1999, those factors were on full display in front of ninety thousand people at the Rose Bowl. Scurry scowled, zeroed in on a single panel of the soccer ball, and made a critical save during the overtime shootout, a deflection that set up Brandi Chastain's winning penalty kick and the U.S. National Team's victory in the Women's World Cup final against China.

Unlike a torn ligament or broken bone, where there's a visible physical injury to assess, it's not always clear what's going on in the brain. After the April 2010 incident, Scurry felt disconnected from herself, like she had been unplugged from a power outlet. Her collision with

Sanderson didn't look like a big deal. It wasn't a huge crash. The referee didn't call a foul. Yet, for three years, she woke up every morning with a low level of pain and white noise in her head that got progressively louder throughout the day, sapping her strength. It took all her will just to take a slow walk outside, the one thing she committed to doing each day. She couldn't sleep or remember things. Scurry describes this time like being "in a hole by myself with no rope" to get out. It left her wondering: Why wasn't she getting better? What was wrong with her? Would she ever get back to being Briana Scurry? She sank into a deep depression with thoughts of ending her life.

It was only after her injury that Scurry learned how many women suffer from concussions in soccer. "I was playing on the best team in the world for many, many years and didn't know that about my own game," she says. Scurry had no idea that when she punted a ball and a midfielder like Cindy Parlow Cone headed it, it was like hitting Cone in the head with a brick. "That was literally what was happening, and no one understood that," she says.

Ten years before Scurry's head injury, Tracey Covassin started asking questions about sex differences in sports-related concussion as part of her PhD research at Temple University. As a Canadian and die-hard hockey fan, she wanted to know more about the injury so prominent in the sport she loved. When she reviewed the existing literature on the topic, she noticed that the studies mostly involved men in ice hockey, football, or boxing. Though some researchers had begun looking into concussion in soccer, studies on women were sparse.

Covassin wondered if the prevalence of concussion in women and men was the same or different. Using the NCAA Injury Surveillance System (now known as the NCAA Injury Surveillance Program), she gathered information on athlete injuries between 1997 and 2000. She narrowed in on sports with both a women's and men's team and that played with similar equipment—specifically soccer, baseball and soft-ball, lacrosse, basketball, and gymnastics. With little precedent to go on, Covassin wasn't sure what the data would reveal. To her surprise, she found that women had a 1.5 to 2 times greater risk for concussion compared to players on the men's teams, particularly in soccer and bas-ketball. Across all sports, women athletes were more likely to sustain a concussion in a game situation compared to their male counterparts.

Covassin, now a professor at Michigan State University, says she didn't realize the importance of her work at the time. "I was just a PhD student. I didn't really know that much back then," she told me, half jokingly. But her study cracked open a new set of research questions and upended long-standing assumptions about who gets head injuries. She explained to me that generally there are more total concussions in football in part because their rosters are bigger compared to other sports. But when Covassin and her colleagues broke down the numbers and calculated injury rates—the total number of injuries in a sport di-vided by the total number of times athletes are exposed to a potential injury in practice or competition—it turned out that the injury rate among women's soccer players wasn't much different from football play-ers. There were 6.71 concussions per 10,000 athlete exposures in foot-ball compared to 6.31 in women's soccer. Subsequent studies confirmed Covassin's findings at both the collegiate and high school levels.

Covassin's discovery echoed what others found in the 1980s and 1990s, when doctors and researchers noticed that women seemed to hurt their knees more compared to men when playing sports with the same jumping, cutting, and pivoting movements. According to the data on NCAA soccer and basketball players, it wasn't just a hunch. Between 1989 and 1999, the rate of injury to the anterior cruciate ligament (ACL)—the ligament that connects the thigh bone to the shinbone and stabilizes the knee—in women's soccer was more than double that found in men's soccer. It was even worse in women's basketball, where ACL injuries were three to four times more common compared to the men's game. Researchers have described the rise of traumatic knee injuries in women's sports as "almost epidemic."

It became clear that a certain subset of injuries clustered around girls and women. Concussions and ACL injuries sounded the loudest alarms: women weren't just getting hurt more often—they also experienced worse outcomes and took longer to recover. With an injury like an ACL tear, most athletes—women and men—opt for surgery and need up to a year for rehabilitation, removing them from training and competition for a significant length of time. However, even when women's knees heal, they are less likely to return to sports or to return at the same level they previously played. They are 15 percent more likely to require a second surgery to repair the ACL again and more likely to sustain a subsequent ACL injury. Men seem to return to sports faster too. For instance, within three weeks of sustaining a concussion, 91 percent of men hockey players were asymptomatic and allowed back on the ice, whereas only 45 percent of injured women players were cleared to return within the same time frame. For some women, concussion symptoms are persistent, lasting six months or more.

In Briana Scurry's case, both she and the medical staff assumed her brain would recover on its own, but it didn't reboot this time. Doctors kept telling her she was within the normal range of healthy, but Scurry wanted to know, "Whose normal?" There was no baseline or standard appropriate for someone like her, a high-performing athlete who had honed her mental skills over years of playing elite sports. "Why are you trying to tell me I'm okay when I'm clearly not okay?" she remembers thinking. With each passing day, Scurry lost valuable time to reconstruct damaged neural pathways and was plagued with prolonged concussion symptoms.

Katherine Snedaker, a social worker in Connecticut, saw a similar dynamic at play among teens who struggled with lingering concussion symptoms. In 2012, she organized a support group and had fifteen kids on her roster—two boys and thirteen girls. However, by the time the group started, both boys had returned to school, their symptoms resolved, and Snedaker was left with an all-girls group. She had prepared talking points for the sessions, but she told me she hadn't needed them. Cuddled on the floor in blankets like a slumber party, the girls just talked. "It was like they were all on a desert island, on different beachheads, and they came up and realized they weren't alone on the island," she says. It felt like that postgame camaraderie found on the bus ride home from an away game. Since people didn't always believe their experience of injury, it was cathartic for the girls to be around others who understood what they were going through.

"I've talked to thousands of women at this point over the last ten years and I basically tell them they're not crazy, stupid, or lying," says Snedaker, who has since founded PINK Concussions, a nonprofit that advocates for sex-specific and gender-responsive strategies for girls and women with brain injuries. "Women have long been the invisible patients of brain injury."

Researchers and practitioners continue to develop models and protocols for injury diagnosis, prevention, treatment, and return to play based on what's observed in men and apply those concepts universally across active populations, but this practice falls short when it comes to women's health. Seventeen years after Covassin's pioneering research and subsequent studies suggested that there were clear and real sex differences in concussion, women are still left out of the conversation. The 2017 Consensus Statement on Concussion in Sport mentioned "sex" only once when referring to potential factors that may predict sports-related concussion or influence its recovery; "gender" is referred to in one footnote. Despite the surge in education and awareness, particularly around injuries like concussion and traumatic knee injuries, the prevalence rates in girls and women haven't improved. Over a twenty-one-year period, the incidence of ACL tears among women has remained unchanged, while it has decreased in men.

More than ten years ago, the lack of information on brain injuries in women sent Scurry on multiple detours on her way to recovery. Despite all the protocols in professional soccer, despite her connections to U.S. Soccer and the U.S. Olympic and Paralympic Committee, Scurry slipped through the cracks. She says, "I should have been able to, at my fingertips, find the right people quickly. It's because I couldn't find them quickly, that's why things got so bad for me." Her battle to find answers morphed into a battle with her health insurance company, which caused further delays.

It wasn't until Scurry met neurologist Kevin Crutchfield at Sinai Hospital in Baltimore that she felt a glimmer of hope. "He saw my eyes and he knew I was broken," Scurry recalls of her first appointment in 2013, three years after her injury. Crutchfield believed the blow to Scurry's temple whipped her head back and damaged the occipital nerve, which runs up the spine and behind the ear. Because it was left untreated for years, the nerve became increasingly muddled, contributing to her headaches. In October 2013, Scurry underwent bilateral occipital nerve release surgery, where surgeons pulled away clumps of damaged tissue that compressed her nerves. When she woke up in recovery, she cried tears of relief because "someone had cut the pain off." With the headaches resolved, Scurry could finally turn to her other symptoms—her balance, her memory, her cognitive processing. A year later, she finally "graduated" from therapy.

When I asked Scurry about the consequences of not studying injury in women at the same rate and to the same degree as men, she told me, "I'm the consequence." By investigating sex-specific differences and understanding women athletes better, we can develop more nuanced guidelines that can optimize their health, recovery, and experience in sports. "It's time. People need to know," Scurry says.

L eading up to the 2012 Summer Olympics in London, Hockey Australia realized they might have a big problem on their hands. Among the squad of thirty-three players on the Hockeyroos, the Australian women's national field hockey team, four players had blown out their knee. It's a pretty high percentage for a sport like field hockey, which tends to have a lower ACL injury rate compared to basketball, soccer, or netball. Hockey Australia wanted to figure out what was happening, why it was happening, and what they could do about it. That's when they met Gillian Weir.

At the time, Weir was working toward her PhD in biomechanics at the University of Western Australia and was interested in injury prevention and rehabilitation. During her time as a strength and conditioning coach for an Australian rules football club, Weir witnessed the fallout from ACL tears. She knew the devastating impact of a traumatic knee injury and the arduous path to recovery. If she could prevent these injuries in some way, she wanted to help. So when Hockey Australia and the Hockeyroos approached the university, it was good timing. Weir set to work with the Hockeyroos to understand the risk factors, develop tools that could identify athletes at greatest risk of ligament trouble, and use that information to head off future injury.

To understand how injuries happen, researchers begin by evaluating the elements that might render a person more vulnerable. There are risk factors external to an athlete, like the rules and equipment used in a given sport, weather, the playing surface (like a grass or turf field), and opponent behavior. Then there are risk factors innate to an athlete, like age, injury history, skeletal structure, joint laxity, genetics, hormones, body movement patterns, and muscle strength. Researchers sort these items into different camps: those facets that can be changed, and those that can't.

In theory, ACL injuries should be preventable. The majority of ruptures happen when someone lands a little off-balance, or their leg does a weird movement when they juke a defender; they aren't the result of direct contact like crashing into equipment or another player. It's the combination of planting the foot and changing direction or speed that sends huge stress through the joint. It implies that there may be something in the way a person moves that puts the knee in a risky position. If that's the case, "we can change biomechanics," Weir says.

Weir, now a senior biomechanist with the New York Yankees, told me it's taken a while for researchers to figure out exactly what combination of forces can cause the ACL to rupture. They can't directly measure ACL strain in a real person, as that would require inserting something like a bobby pin into the knee and wrapping it around the ligament—not exactly a comfortable proposition. Instead, researchers study people in the lab, examining their movement patterns as they complete a task like a single-leg jump or a sidestep. They've analyzed videos to understand athlete kinematics during game situations, rigged up knees from cadavers and applied different forces to the tibia and femur to determine what's required to strain the ACL, and used computers to simulate the application of different amounts of force to the side of the knee to see what happens. From these studies, researchers identified the torques, or stresses on the knee, that elevate ACL strain the most.

It turns out that when women land or sidestep, they tend to recruit the quadriceps muscles in the front of the thigh to stabilize the knee, as opposed to relying on the multiple muscles in the back of the body, like the hamstrings, glutes, and calves. When the quads are engaged, the knee is pulled back into a straight, extended position. When a person lands with a straighter knee, the ground reaction force slaps the knee back even farther, channeling more stress through the joint and ACL. In some of Weir's studies, she also noticed that hip mechanics of women and men weren't the same when they had to sidestep suddenly. "Females have poorer hip control, so they internally rotate and then their knee buckles in more than males," she says, a classic ACL injury red flag.

Variations in biomechanics offer a reasonable explanation for why women may become injured more frequently, but those movement patterns don't exist in a vacuum. There are numerous factors that come into play to create the chain reaction that leads to real-life injury. To tease apart the equation, in the past, researchers often homed in on the ways in which women's and men's bodies were dissimilar. Intentional or not, this framework drew on an old narrative: since men were considered more durable and injury-proof, there must be something inherently different (and wrong) about women's bodies that makes them more injury-prone.

Intuitively, it seemed to make sense. Injury rates between girls and boys start to deviate at puberty, right when young bodies separate developmentally. The divergence in body size and strength along with differences in hormones seemed to be likely suspects behind the rise in injuries. For example, neck strength and girth can influence how well a person can stabilize their head during a collision or other knock to the body. During a collision, women experience higher acceleration and rotational forces potentially due to their thinner necks, lower levels of neck strength, and smaller heads. And when it comes to concussion, those forces matter.

When the body is hit externally—whether it's direct head-to-head contact, the head smacking the ground or a soccer ball, or even a jolt to the torso—brain tissue rapidly deforms, like Jell-O. The tissue and microscopic neurons (the nerve cells that act as the brain's electric grid) twist, shear, and bend. In some cases, the neurons escape unscathed. In others, they are damaged and effectively taken offline. As Doug Smith, a neurosurgeon and professor at the University of Pennsylvania, explained to me, a concussion that leaves someone dazed and confused is like a brownout of the city. If they lose consciousness, he says, "it's like a blackout where the whole network is taken down."

Researchers had a hunch that concussion had something to do with axons, the long, thin structures that extend from the main part of the nerve cells. They pass messages along tiny train tracks called microtubules. Under normal conditions, axons are flexible. The microtubules, which are stabilized by tau proteins, slide past one another when the axon moves. But when axons are stretched rapidly, like during a hit to the head, the tau proteins can't unfurl fast enough to keep up with the movement, and the microtubules rupture. "Your brain can literally break down at the nanoscale," Smith says. The protein transported along the microtubules then derails and piles up, causing swelling along the axons. What's more, the damage disrupts the brain's electrical signaling. The channels that regulate the precise flow of sodium ions, needed to create a spark of electricity, are stuck open like a flap valve in a toilet tank. Too much sodium floods the system, and the brain can't send a signal down the axon. When neurons don't work efficiently, everything slows down—reaction time, awareness, attention—the short- and long-term symptoms people experience after a head injury.

Since the same injury mechanism appears to be at play in both women and men, Smith and his colleagues were curious whether axon size differed between the two groups and influenced concussion outcomes. In the lab, they created "mini-brains" that were either genetically female or male. When they stretched and applied the same mechanical force to both types of axons, similar to what's seen with concussion, they saw a dramatic difference. In female axons, more microtubules broke, and twenty-four hours after injury, there was greater swelling and disrupted electrical signaling compared to the male axons. Researchers believe that the male axons' larger diameter and sturdier architecture may protect them from more harmful injury. But Smith points out that it may just be an issue of size, not necessarily sex. Smaller male axons may exhibit the same vulnerability, but right now, they don't know.

In recent years, researchers have started to pay more attention to hormones and how they might influence physiology and injury risk. Estrogen has received the most attention since there are estrogen receptors present in bone, muscle, and connective tissues. When engineered ligaments were exposed to the hormone, it inhibited the activity of lysyl oxidase, an enzyme that's integral for creating cross-linkages between the collagen fibers that stabilize the tissue. With fewer cross-links, the ligament was more pliable, weaker, and more prone to fail under stress, even after only forty-eight hours of estrogen exposure. Theoretically, this could mean that the normal rise and fall of hormones seen in menstruating people affects ligament stability and ACL injury risk. And it does appear that knees become more lax over the course of the menstrual cycle, peaking during the late follicular and ovulatory phases when estrogen surges. People who don't menstruate, on the other hand, don't experience changes in knee laxity over time. If hormonal ebbs and flows are associated with greater ACL injury risk, researchers wondered if hormonal contraceptives and their steady supply of estrogen would provide protection. The answer, as with much of this research, is maybe.

While high levels of estrogen may be a bad thing for ligament function, it appears to bolster injury resistance in tendons, the tissue that connects muscle to bone. When a muscle is under tension, a flexible tendon helps dissipate the load by absorbing some of the stress. A stiff tendon isn't as resilient.

Premenopausal people, generally, are at a lower risk of an Achilles tendon rupture compared to men. The situation flips after menopause, when estrogen levels are low and the risk for tendon injuries becomes more comparable to men's. Interestingly, the decline in estrogen means postmenopausal people are less likely to tear their ACL.

While focusing on intrinsic risk factors is a convenient framework to make sense of what clinicians and researchers witnessed in their clinics and labs, "it's a kind of chicken-or-the-egg scenario," Weir says. Do women and men experience different injury risks because of sex and biology or because of gender? Or both? If women are repeatedly told their bodies are faulty and more prone to fail, does the psychological burden of that negative stereotype lead to a self-fulfilling prophecy? Women may also be less susceptible to muscle injuries. For instance, among elite women and men athletes, women suffered 52 percent fewer muscle injuries.

As a high school student at Rowmark Ski Academy in Salt Lake City, Utah, Lauren Samuels remembers when her coach asked the team to read an article detailing the relationship between the menstrual cycle and knee injury. "It was so negative," the former U.S. Ski Team member told me. "We're in an incredibly dangerous sport with high injury rates in the first place, especially ACLs, and then you just put this on top of us? What do you want us to do?" Sure, she and her teammates could go on hormonal birth control, but there still wasn't much research to support the efficacy of oral contraception to protect against knee injury. They could try to time their cycles so they didn't coincide with peak racing season, but trying to convince one's hormones to behave in an orderly fashion, let alone syncing up a whole team to an advantageous schedule, seemed farcical. There weren't any viable options. She felt like her coach was telling the team they'd inevitably get hurt because of their cycles, an inherent part of their physiology.

When women do get injured, the emotional trauma can linger, taking a toll on the mind as well as the body. While both women and men were less confident and hesitant to return to sports following an ACL reconstruction, women tend to endure greater anxiety, fear, and mental anguish after the injury, including symptoms similar to post-traumatic stress disorder. It's hard not to be conscious of a maimed joint or question whether it will hold up when tested during rigorous physical activity. Among recreational women athletes who previously injured their knee, more than 70 percent changed their training practices and more than 31 percent avoided specific movement patterns to prevent "wear and tear," for fear of hurting their knees again. Given that injuries like ACL ruptures can happen in an instant, lack of confidence or nagging worries can influence muscle recruitment and alter body mechanics. Tellingly, those who were most afraid of reinjury were thirteen times more likely to experience a subsequent ACL injury within two years.

In recent years, researchers have started to push back on the scientific establishment and its propensity to turn to biological explanations for injury. The current paradigm "displaces bodies from the conditions of our existence, instead focusing heavily on biomechanics explanations, as if our muscles and joints are not impacted by the weight of our life experiences," writes assistant professor of physical therapy Joanne Parsons and colleagues in a recent paper in the *British Journal of Sports Medicine*. Doing so implies that women are destined to suffer knee, head, bone, and other injuries just because of their hormones and physiology.

Rather, environmental factors like exposure to sports during childhood; training and competition opportunities; and access to facilities, coaching, and medical staff—and the gendered nature of these elements— all intersect and drive injury. Gillian Weir told me this was a major topic of discussion in Australia after the 2017 debut of Australian Football League Women's (AFLW), the women's elite league for Australian rules football. In the first two seasons, players on the women's side were at least six times more likely to tear their ACL compared to their male counterparts.

On the surface, it seemed like the typical ACL injury story in women athletes, where the ligament ruptured when someone landed on one leg or sidestepped to change directions. But zooming out, another picture came into focus. Historically, men had more opportunities to compete at higher levels of Australian rules football, but the sport was still in its infancy among women. Many AFLW players weren't exposed to the game as kids and missed the critical developmental period men were privy to, leaving them with a shorter, steeper on-ramp to develop the sports-specific skills and physical conditioning necessary for elite competition. AFLW teams were also encouraged to recruit rookies, athletes who hadn't played Australian rules football in at least the past three years. For example, former Olympic javelin thrower Kim Mickle played for the Fremantle Dockers. While she had played "footy" at the junior level, she devoted her elite athletic career to track and field. Training for and throwing a javelin is very different from playing a dynamic team sport where you jump, run, kick, and change direction, all while keeping track of your teammates and opponents. Mickle tore her ACL in her first AFLW game.

Increasing evidence suggests that early experience and exposure to movement patterns matters. In dance, a discipline where girls and women far outnumber boys and men, there's not the same gender-based disparity in ACL injuries, even though dancers perform powerful leaps and often land on one leg. In part, it may be because girl and boy dancers learn jumping, landing, and balance technique at a young age and continually practice these skills throughout their dance careers.

To test this hypothesis, researchers compared the landing mechanics of women and men who were professional modern and ballet dancers. Participants stood on a 30-centimeter platform on their dominant leg with their nondominant leg lifted. They crossed their arms over the chest, then dropped off the platform and landed on a force plate, which is like a bathroom scale. In a separate study, using the same single-leg drop test, researchers observed dancers and team sports athletes. The women and men who danced and the men who played team sports all used similar landing strategies, especially those who started training at an earlier age, and minimized excessive movement at the knee and hip. When women team sport athletes landed, their knees wobbled and collapsed inward, a harbinger of ACL injury. In other words, faulty biomechanics aren't inherent to women. It depends on learning good technique, particularly at a young age, and repeated opportunities to practice and ingrain those neuromuscular pathways in the body.

Similarly, strength is often identified as a clear sex-related injury risk factor, since men generally have greater muscle mass than women. But whether or not an athlete is taught and encouraged to lift can be influenced profoundly by gender norms. Muscularity has long been a way for men to visibly demonstrate their manhood. Women are often wary of becoming too muscular, fearing it would be considered crossing the line into masculinity, even as more and more women took up sports in the twentieth century. Not only do these stereotypes continue to discourage girls and women from lifting weights (and getting too bulky), they delineate the physical space in gyms too. Historically, weight rooms are the domain of men, while women are relegated to the treadmills, elliptical machines, and group fitness studios.

A survey of high school varsity coaches found that among coaches working with boys, half required athletes to strength train—only 9 percent of the girls' coaches required the same. Coaches were also more likely to ask boys to strength train three or more days per week, as well as year-round. The girls' programs, instead, were more likely to include exercises considered more "feminine," like yoga, Pilates, and low-intensity, high-repetition routines. As a result, girls and women may not take up strength training until a much later age.

Eliza Stone, a top-ranked saber fencer for Team USA, told me she never learned how to strengthen her body to make it more resilient. "We look like crabs moving up and down [the strip] in that en garde position, and then lunge on one side," she says. The body ends up lop-sided from all the single-sided work and can take only so many lunges before overuse injuries crop up. Women fencers, in particular, experience more hip and knee injuries across all weapons in the sport—foil, épée, and saber; Stone has herniated a disk in her back, torn her hip labrum, and has chronic pain in her knee. Yet her coaches never encouraged resistance training, and traditional strength and conditioning coaches didn't always know how to work with Stone or understand the demands of her sport. "They trained me like they train their college bros," Stone says. Or they sent her to the corner to stretch.

Weir told me that in the case of the Australian rules football leagues, the men's teams are more likely to have a stacked support team, in-cluding high-performance coaches, sports science analysts, and strength and conditioning coaches. Women's teams, on the other hand, may have one strength and conditioning coach, who is likely paid a fraction of what the men's team coach earns. The discrepancy in pay can subse-quently influence the caliber of candidates who apply for the position, with more qualified candidates vying for the higher-paid positions with the men's side. While AFLW player salaries are proportional to those in the men's league, the women's season is shorter, which translates to a sizable pay gap. Athletes on the women's side are more likely to need part-time jobs to supplement their pay, cutting into the time and energy available to train for their sport.

Christina Master, a sports medicine physician and concussion specialist at Children's Hospital of Philadelphia, believes that unequal access to resources plays a greater role in what are thought to be non-modifiable injury risk factors, more than clinicians and researchers have realized. Master was part of a research team that looked at sports-related concussion in girls and boys between the ages of seven and eighteen. As suspected, girls took longer to heal, return to school, and get back to sports. But, compared to boys, more time elapsed between a girl's injury and her concussion evaluation by a medical specialist. If girls and boys saw a doctor within the same time frame, the differences disappeared; girls and boys recovered at the same rate.

Master found something similar at the collegiate level as part of the Concussion Assessment, Research and Education (CARE) Consortium, the largest collegiate study on concussion. Sifting through data on more than one thousand concussions in both women's and men's sports from thirty institutions, the team found there was no significant difference in recovery or return-to-sports rates. However, when they segmented the data, women took longer to recuperate in contact sports compared to men. And men participating in limited-contact sports, like gymnastics, took longer to bounce back than women. The researchers looked at the data by division level too. There were no differences in recovery rates at the Division I level, but women competing at Division II and III schools took longer to get better than men.

If biology alone drove disparities in concussion recovery, the sex-based differences should hold up regardless of when someone saw a doctor, the type of sport, or NCAA division level. Master believes the discrepancy potentially comes down to resource allocation. At schools with limited resources, like Division II and III schools, athletic trainers may be stationed on the sidelines of men's games, but not women's. Division I schools may have more available staff to go around. When it comes to contact sports, again, resources may be distributed based on an athletic department's priorities, with men's contact sports getting first dibs over the women's teams and sports like men's gymnastics.

From this standpoint, external factors and the environment around women's sports aren't always set up to get women care quickly when they get injured. The Football Association, the governing body for soccer in England, mandates different levels of medical support between the men's Premier League and the Women's Super League. On the men's side, four medical doctors are required at every match—a club doctor from both the home and away teams, a medical coordinator or "tunnel doctor" who serves as an extra pair of eyes to identify potential concussions via replays, and a "crowd doctor." Women's teams are only mandated to have one doctor on-site during matches, the home team physician. The medical team doesn't have access to real-time video replays, which could delay identification and assessment of potential head injuries, greatly compromising the safety and well-being of women players.

The traditional injury model is an oversimplification of a complex story. For injuries like sports-related concussion, ACL tears, and stress fractures, most non-modifiable risk factors go hand in hand with women's physiology—wider pelvis, hormones, lax joints, and fragile frames. But focusing on biologically based sex differences perpetuates the idea that there is something inherently defective and risky about women's bodies and their participation in sports. It can be a particularly disempowering narrative, leaving girls and women in a seemingly helpless situation. There's not much they can do to change their physiology, anatomy, genetics, or hormones. "It's important to understand what's going on, but the overemphasis [on biology alone] can also be dangerous," says Monica Rho, head team physician for the U.S. women's national soccer team and associate professor at Northwestern University. "The research that goes into women in sports medicine should not be utilized or interpreted to limit one's ability to train or perform."

We'll never eliminate injuries completely from sports—it's one of the risks you assume when you lace up your sneakers, clip into your bike, or pick up a racket, ball, or bat. But if clinicians and scientists want to change the trajectory among women athletes, they need to reimagine the injury model and consider athletes within the context of their entire lived experience. Looking at education, access, resources, exposure to sports, and training alongside physiology, anatomy, and hormones is crucial. Considering these factors together creates a more inclusive model of injury that moves beyond just the binary sex of male and female.

On the injury prevention front, clinicians and sports science practitioners need to do more than just screen for risk. They need to provide girls and women with the tools to actually *reduce* those vulnerabilities. That's what Gillian Weir hoped to do when she designed the ACL injury prevention program for the Hockeyroos.

She started by understanding the problematic movement patterns and postures that elevate ACL stress and then worked backward to identify appropriate countermeasures. From the research on biomechanics, Weir identified four areas where improved movement patterns could decrease knee strain—increasing knee flexion at foot strike so you don't land in a straight-knee position, improving trunk and upper body control, strengthening hip muscles to keep the knee from buckling in, and increasing calf muscle strength. She compiled a list of exercises that targeted each of these areas and handed it off to Kate Starre, who was then the Hockeyroos' strength and conditioning coach.

Starting with the 2013–2014 season, the Hockeyroos implemented a nine-week intensive program—four twenty-minute sessions per week integrated into their warm-ups and strength sessions. Starre would pick and choose exercises from Weir's list that complemented the team's training priorities. After the initial nine weeks, the team continued the program for another sixteen weeks, scaling down the sessions to three ten-minute blocks per week in order to maintain and consolidate the work from the previous phase. They continued the program through the 2014–2015 season too.

During each phase of the study, Weir tested the athletes and assessed their injury risk in the lab by having them run at a set speed toward a TV situated 15 to 20 meters ahead. When an arrow appeared on the screen, the athlete quickly pivoted and ran in that direction. When the athlete landed on the force plate in the middle of the room, it told researchers how much load traveled through the knee. Weir points out that it's important that the change in direction was unanticipated. When athletes don't have time to plan where they're going, the stress in the knee is twice as high compared to when they have more time to prepare.

At the end of the two years, there were no ACL injuries among the twenty-six Hockeyroos players, and athletes lowered potentially dangerous strain in their knees. In particular, those who were considered high-risk for ACL injuries reduced their peak knee valgus moments (the knee buckling action) by 30 percent and improved the way they activated their muscles. Players also maintained or improved their overall strength, speed, and aerobic power. Overall, ACL injury prevention programs like Weir's have been successful, lowering the risk of injuries by half in all athletes, and non-contact injuries by two-thirds in women athletes.

Greg Myer, director of the Emory Sports Performance and Research Center, has taken injury prevention a step further by incorporating augmented reality. Using three-dimensional motion capture analysis, Myer and his colleagues collect information on how an athlete moves when jumping, squatting, and cutting to identify biomechanical risk factors. Then the athlete dons an augmented reality headset. It shows them a simple geometric shape that represents the athlete's movement patterns. It moves and shifts as the athlete does. Athletes are told to keep the shape as close to a rectangle as possible when performing a series of exercises. In doing so, the shape helps the athlete avoid high-risk biomechanics and provides real-time feedback.

The protocol is not only research driven but also highly personalized to the individual by focusing on the mechanics of how the whole body moves. Since the program is self-guided, athletes are forced to problem-solve on the spot and figure out how to avoid tricky movement patterns, which Myer says hardwires the information into their brain. He likens the program to an MRI, but one that does surgery in addition to scanning the body. "I think that's the future of training," he says.

In an effort to improve diagnosis and treatment, clinicians, health-care providers, and researchers have also developed multidisciplinary models for clinical care *specifically* for active girls and women. In 1997, orthopedic surgeon Jo Hannafin and sports medicine physician Lisa Callahan founded the Women's Sports Medicine Center at the Hospital for Special Surgery in New York City, the first multidisciplinary center focused on the needs of women athletes. More programs have emerged in recent years, including the Female Athlete Program at Boston Children's Hospital, the Center for Women's Sports Health at NYU Langone, and the Female Athlete Science and Translational Research (FASTR) Program at Stanford University.

Kate Ackerman, director of Boston Children's Hospital's program, says, "Most sports practices in the country were orthopedics-based and a lot of athletes were getting siloed treatment," with injuries treated as purely physical. Prior to founding the program in 2013, Ackerman might have seen a patient with multiple bone stress injuries through her sports medicine practice. As they talked, she'd learn the woman was being treated for an eating disorder by a different practice and for irregular periods by her gynecologist. But no one talked to one another or connected the dots. Often, the information and care provided weren't tailored to someone who identified as an athlete or active person and the specific needs that lifestyle presents.

What's different about programs like Ackerman's is the degree of interdepartmental collaboration. Since injuries can be intimately linked with nutrition, mental health, and hormonal health, she brings together surgeons, physicians, nutritionists, physical therapists, exercise physiologists, and sports psychologists to provide comprehensive care under one roof. In doing so, they address the underlying behaviors and sociocultural issues that might drive injuries, while also recognizing the girls and women as whole people.

That means someone with recurrent stress fractures isn't just a collection of blood test results and bone density scans. They are connected with a sports medicine doctor or surgeon who asks questions about hormones and menstrual cycles, a registered dietitian who understands the nutritional demands of an active person, and a mental health professional who can address any underlying disordered eating issues—all critical aspects needed to cultivate not just bone health, but health overall. An athlete with a torn ACL isn't just treated by an orthopedic surgeon and sent off to physical therapy; they may also meet with a mental health professional to address the anxiety and concerns that come along with injury and help nurture the psychological readiness and confidence that's essential to returning to sports.

But in order for healthcare professionals to continue to provide the best care, they need more research on girls and women in sports—research that understands the nuances of women's physiology, that asks the right questions about the mechanism of injury in women's bodies and the rehabilitation protocols needed to return people to physical activity and to prevent future injury. For instance, according to Ackerman, most clinicians try to get people back to activities like running by twelve weeks after a stress fracture diagnosis. But when she and her colleagues looked at bone density and quality during that time frame, they were surprised to learn that the bone microarchitecture actually got worse. These findings suggest that the bone needed more time to recover. There's a clear opportunity to better understand injury and rehabilitation better so that people can get back to their activities while avoiding further injury.

Most importantly, we need to untangle the upstream effects of environment, culture, and social and gender norms on sports and provide more equitable access to funding, facilities, and resources. "If women have equal access to all educational resources, including sports and its attendant athletic training and sports medical care, that can basically eliminate disparities that are potentially biologically based and affect outcomes," Christina Master says.

More and better resources are an important step toward leveling the playing field for girls and women and eliminating conditions that make them more vulnerable to injury, but it's not the only step. For girls and women to play in the first place, they need athletic clothing, attire, and gear suited to their bodies, and that starts with the sports bra.

7.

Bounce Control

· · · ● ● ● ● · ·

There has never been a good solution for supporting breasts during physical activity. Ancient Greek and Roman women wrapped their breasts in wool and linen. In the Victorian era, tennis players at Wimbledon returned to the dressing room between matches to remove bloodied corsets, the result of the metal and whalebone stays gouging their skin when they ran, lunged, and swung at the ball. In the late 1890s, sports corsets, like the Ferris's Good Sense Corset Waist, were introduced to better suit activities like bicycling, tennis, and riding. With elastic slides instead of hard steel and bone stays and buttons in place of metal clasps, advertisements described the resulting garment as yielding "to every motion of the body, permitting full expansion of the lungs, at the same time giving the body healthful and graceful support." Nonetheless, it was still a corset.

As the women's liberation movement swept across the United States in the 1970s, a revolution took hold in sports and fitness, challenging the idea that women weren't cut out for physical activity. Millions laced up their sneakers for the first time during the "jogging boom," and Title IX ushered more girls and women onto the playing field and starting lines across the country. But there still wasn't a good sports-specific bra to support and control breast movement.

Among those to take up running in the late seventies was Lisa Lindahl. She was in her twenties and worked part-time at the University of Vermont (UVM) in Burlington while she finished her undergraduate degree. The first time she ran, she couldn't make it around UVM's indoor track—one-tenth of a mile—without stopping. When she finally ran a mile, she felt like she'd won an Olympic medal. She described running as a "trance dance." The *whomp-whomp-whomp* of her feet on the ground and the rhythm of her breath energized her and cleared her mind.

But Lindahl couldn't figure out how to stop her breasts from bouncing when she ran. When she searched for an athletic brassiere at the store, salespeople stared at her blankly. With no alternative, she donned a regular bra on her jogs, usually one that was a cup size smaller than her normal size. The nylon material rubbed her skin raw, the straps slipped off her shoulders, and fasteners and hardware scraped her skin. She tried running without a bra, but that wasn't exactly comfortable either.

Lindahl wasn't the only one to struggle with jiggling boobs. Her sister, also new to running, wondered why there wasn't something akin to a jockstrap for breasts. When she posed the question to Lindahl over the phone, the siblings broke down in peals of laughter—but Lindahl recognized a kernel of an idea. She grabbed a notebook and scribbled down the requirements for such an undergarment. It had to be made from breathable material that minimized sweat and didn't chafe. It had to be free of hardware that might bruise the ribs. It had to have straps that stayed put. Most importantly, it had to control breast movement.

Lindahl convinced her childhood friend Polly Smith to help her come up with a workable prototype. Smith was a costume designer, and initially she thought Lindahl was crazy. Smith told her that the only thing more difficult than designing a bra was designing a shoe. It was more an engineering task than a design one.

Developing a decent sports bra proved to be challenging. Smith cut apart everyday brassieres and sewed them back together, but nothing worked quite right when Lindahl tested the prototypes. One day, while the women were at Lindahl's house, Lindahl's then husband sauntered down the stairs with a jockstrap pulled over his chest. Jokingly, he said, "Here ladies. Here's your jock bra!" Giggling, Lindahl jumped up and took the jockstrap from him, pulling the waistband around her ribs and stretching the pouch over one of her breasts. It took her a moment, but she noticed it felt right. Her breast felt supported.

The next day, Smith sent Hinda Miller, who worked with her at the Champlain Shakespeare Festival in town, to buy two jockstraps. They then transformed the wide waistband into an underband and crossed the straps in the back. The two pouches became the bra's cups, and the compression fabric pulled the breasts closer to the body, minimizing the up-and-down breast motion. Lindahl says the bra looked like a "frankenbra," but it worked.

In a way, being outsiders to the bra industry was an advantage. For Smith and Miller, the creativity and innovation inherent to costume design liberated them. "When you design for a show, you do whatever it takes to fulfill the director's vision," Miller told me. They didn't think twice about sewing two jockstraps together, something traditional bra designers might have scoffed at. At the end of the summer, Smith returned to New York City to pursue her career (she went on to win seven Emmys as a costume designer for the Jim Henson Company) while Lindahl and Miller tried to figure out how to manufacture and sell their sports bra.

Initially, things didn't run smoothly. Miller told me she and Lindahl didn't know the first thing about the retail industry, manufacturing, marketing, and sales cycles. The product was originally named the Jock

Bra, but upon learning that women didn't like to be called jocks, they changed it to Jogbra. They wanted women to buy their sports bra where they bought their running shoes, but specialty running stores weren't used to selling undergarments. Miller and Lindahl needed to develop packaging and language to describe their product so that running store employees, who were mostly men, would feel comfortable promoting the athletic brassieres. Despite turning a profit the first year, sale projections and inventory management became more complicated as sales went up and their product line expanded. There was no automated algorithm to guide them. It was the epitome of on-the-job learning, but they soon worked out the kinks.

The emergence of the sports bra five years after the passage of Title IX was a significant and monumental milestone for active women, and the two are inextricably linked. Lindahl, Smith, and Miller created a completely unique product category, one that wasn't a scaled down version of men's gear. This singular garment enabled women and girls to participate in sports—whether it was running, hiking, yoga, or skiing—in greater comfort and with less embarrassment.

Jogbra, with its byline "By women, for women," became a juggernaut, dominating the marketplace for years. The original model, preserved in the Smithsonian Institution, evolved into a line of low-impact, medium-impact, and high-impact sports bras. In 1990, Lindahl and Miller sold the company to Playtex, where it became known as Champion Jogbra. As growth and participation in sports and fitness has continued to grow, so has the sports bra market, which recorded $9 billion in revenue in 2019 and is expected to reach $38.4 billion by 2026.

While the advent of sports bras lowered the hurdle to participate in exercise and sports, it's no secret that they aren't the most comfortable or flattering article of clothing. Sure, bra designs have advanced since the Jogbra of the 1970s, but people with breasts still aren't satisfied with the options available. Shopping for a sports-specific bra that fits, works well, and is comfortable can make a woman feel like Goldilocks searching for the perfect bowl of porridge. This one is too tight. This one is too loose. This one doesn't provide enough support. This one's straps slip. This one is a struggle to get on and off. But as anyone with breasts will tell you, sports bras are a necessary evil if you want to be active or play sports—regardless of age, level, or activity.

But sports bras haven't always been given their due respect. They're written off as just a piece of spandex, without recognizing the scope of their immense and multifaceted job. Since breasts are made of soft tissue, they're pretty unruly. When the torso moves, there are no bones or muscles to act as internal scaffolding and prevent breast tissue from being stretched and pulled like Play-Doh. While skin and fibrous tissue called Cooper's ligaments provide slight support, these structures aren't designed for heavy lifting or stabilization. To make matters more complicated, breast size naturally fluctuates as much as a full cup size over the course of a month. For a B cup, it's not a big deal, but it can mean a difference of a full cup size for a DDD. Breast shape and size can also change during different life stages like puberty, pregnancy, postpartum, and menopause.

How much or how fast breasts move depends on cup size and the intensity of the activity, and it's not an insignificant amount. It's a basic physics problem, where force equals mass times acceleration. Bigger breasts accelerate and move faster and more compared to smaller breasts. In one study, breast displacement ranged from 4 centimeters when walking to 15 centimeters when running with no bra on. When people performed jumping jacks, their breasts moved as much as 19 centimeters. It can be uncomfortable, to say the least, and roughly one-third of active people with breasts report breast discomfort with vigorous activity. Among elite athletes, that percentage is even higher;

44 percent said they experienced pain while training and competing. Unsurprisingly, a well-fitting and supportive sports bra reduces the magnitude of movement, the force on breasts, and perception of pain better than everyday bras or a no-bra situation. Not only does a good sports bra minimize discomfort, it can also decrease the likelihood of tissue damage and early onset of breast sagging—no matter one's cup size.

There's also the intangible, emotional role sports bras play. People with breasts aren't just searching for physical support (although that's a priority). They want psychological comfort too, and that can come in different forms.

LaJean Lawson is an exercise scientist and sports bra guru who has consulted for Champion on their sports bras since the 1980s. She says some bra testers told her they don't want a sports bra that smashes them down because they are already small-chested. Others seek modesty. When Lawson asked an elite marathon runner why she chose to wear a sports bra with padding during competition—wouldn't she want the thinnest, lightest garment when racing?—the runner told her if a photographer captures her crossing the finish line with her arms up, the whole world would see her nipples. That's why she turned to padding, especially during races. The feedback prompted Lawson to ask bra testers to rate the "modesty" of the products they tried in her lab, or the extent to which the bras called attention to them in ways they *didn't* want.

It's led to interesting revelations. Some people won't wear sports bras that have a little bump, seam, or zipper that pokes out under their shirt. They told Lawson that they didn't want someone's gaze to linger on their chests while they tried to figure out what the bump was. Others want to feel a certain way in their sports bras to account for how they want to feel internally. "[Sports bras] have the power to fuel their confidence, to make them feel powerful, to make them feel prepared or focused or exactly how they want to feel at any moment in time," Chantelle

Murnaghan, vice president in charge of product innovation at Lululemon told me. "When you feel your best, you perform your best." Former professional runner Lauren Fleshman wrote on Instagram, "Two different sports bras might do the job well on paper, but one wins out when I account for the fact that tempos make me anxious, which makes me more sensitive to being squeezed." To support her desired mindset, Fleshman might choose a bra that makes her feel calm and more confident during a tough workout.

Taken together, sports bras and breast motion can influence athletic performance in very real ways. Anecdotally, people with breasts have talked about this for years. Professional tennis player Simona Halep opted for breast reduction surgery at age seventeen because she felt her larger chest hindered her reaction time. When researchers dug into the biomechanical data, they found there's quite a bit of truth to that perception. People may experience greater ground reaction forces when running in a low-support bra. To compensate for the discomfort, they may change their running gait, particularly those with larger busts. Their legs may not swing through their full range of motion, resulting in a more stilted stride. Some people brought their arms closer to their ribs to control side-to-side breast motion. The upper body, particularly the pectoral and deltoid muscles, has to work harder to maintain that tighter, more restrictive posture, which then costs energy and increases muscle tension. Others breathed less frequently, holding their breath to lessen the pain they experienced with each step.

These compensations are the opposite of the relaxed, energy-efficient running posture and quick turnover experts recommend. When runners wore a poorly fitted sports bra, researchers found that they took steps that were, on average, 4 centimeters shorter compared to when they wore a bra that supported the breasts well. Over the course of the 26.2 miles of a marathon, that translates to a performance dip of approximately 2 percent.

In another instance, Jenny Burbage, a scientist with the Research Group in Breast Health at the University of Portsmouth in the United Kingdom, one of the leading breast biomechanics labs in the world, worked with an elite rower. The athlete struggled with inflamed intercostal muscles between her ribs and she wore a loose compression-style sports bra, even after she recovered. Burbage fitted her for a sports bra and watched her row on an ergometer in the lab. Now the athlete could pull the handle farther back in a long, fluid stroke closer to her body, which is critical for generating power. The bra fit snugly around the rib cage, but it didn't affect the athlete's lung capacity or make it harder to breathe. Burbage thinks the athlete's biomechanics improved because of the way the breasts were positioned in the better-fitting sports bra. While current research certainly does suggest that breast size, movement patterns, and comfort could have an effect on performance, Burbage says more studies are needed to examine this relationship further.

Despite being an essential garment for active people with breasts, finding a good sports bra is fraught with frustrating challenges. It's a highly individual choice—how people want their sports bra to fit and feel, from the amount of motion control to the amount of compression, can differ greatly from one person to the next, even if they wear the exact same size and participate in the same sport. Those preferences also vary by age, breast size, breast tissue composition, and type of physical activity.

In January 2019, more than six hundred people responded to an *Outside* survey about sports bras, representing band sizes between 32 and 48 and cup sizes from AA to I. Sixty-one percent said they weren't happy with the options for high-impact sports bras. Forty percent weren't satisfied with low-impact models either. The biggest complaints? The bras were uncomfortable and didn't look great, and people had to make numerous trade-offs. They have resorted to double- and triple-bagging their breasts (wearing multiple sports bras at one time) and other means to secure their chests and make them conform to what they think is appropriate to pursue their sport. With high-performance sports bras for larger cup sizes virtually impossible to find, the limited selection plays into collective notions of what's considered an athletic body. "If you are not five foot six and a 34B, then you are destined to make weird compromises, or spend an obscene amount of money," says Lorna McLean-Thomas, an active mother from Colorado.

Sports bra sizing is an imprecise system at best. There's an alphabet soup of cup sizes (originally meant to distinguish small from large cups and not aligned with any real volumetric details), numeric underband sizes (a seemingly arbitrary calculation based on the circumference of your rib cage), and small/medium/large sizes that superimpose themselves on top of all those cups and band sizes. Most brands and manufacturers use their own sizing charts and don't always offer a full bevy of sizes in their sports bras, forcing people to squish a wide range of body shapes and proportions into a small number of products. With fewer brick-and-mortar stores and sporting goods retailers, there aren't many opportunities for people to try on garments to see what fits, leaving many people stuck with the wrong size.

Maria Napolitano, a professional in the publishing industry, was an active kid. But when she hit her teens in the mid- to late aughts, she moved away from higher-impact sports like running and softball because she couldn't find a sports bra that fit her chest, which she describes as "pretty busty." Instead, she stuck with fencing, where the chest protector and jacket smushed down and covered up her breasts. "Looking back, I think that was one of the reasons I was more comfortable continuing the sport through high school and college," she told me. Years later, when Napolitano wanted to return to other activities, particularly running, she couldn't find anything that was more supportive than her old stretched-out sports bras. Her proportions didn't match what was available in stores or online. She resorted to wearing two sports bras, which led to chafing. "It was embarrassing," she says.

Ultimately, breasts can keep people from being active and partici-pating in sports. "I have a lot of angst over how many years of my life I wanted to be an active, athletic person and felt like there were obstacles to that," Napolitano says. "I was held back by my own body and the pain from bouncing and jiggling." Napolitano voiced a feeling of shame and frustration that many people experience. When asked to rate barriers to exercise, women between the ages of eighteen and sixty-five listed breasts as the fourth biggest reason they aren't active. They were ranked just behind energy and motivation, time constraints, and health, but ahead of cost and availability of exercise facilities.

If people can't find a sports bra that fits, their breasts won't be sup-ported. When their breasts move around a lot while exercising, it can lead to pain, awkwardness, and embarrassment. And when they're in pain or self-conscious while exercising, they may be more likely to quit—or choose not to participate at all. An ill-fitting or unsupportive sports bra can also lead to other issues, like poor posture and muscular problems because upper body muscles must kick in to do the work your bra should be doing. Straps that leave deep furrow marks in the shoulder can cause nerve pain. It's a vicious circle.

One reason sports bras often don't perform well is because for a long time the science of breast motion was overlooked. Breasts didn't seem worthy of investigation to the bra designers and biomech-anics researchers, who were mostly men, and it's taken a while for the field to gain traction. The field of breast biomechanics largely didn't exist before LaJean Lawson fell into it nearly forty years ago. In the 1980s, she was a graduate student studying historical costume design at

Utah State University, but her research projects kept falling through. A conversation with a physical education professor led Lawson to sports bras. After reviewing the literature, she realized no one had really looked into breast movement and decided to take it on.

In those early days, Lawson encountered her fair share of skeptics. After she and her coinvestigators received a grant for their work, they were awarded a Golden Fleece Award. Lawson told me it's the equivalent of the Razzies, the satirical award that honors the worst films in the industry, but instead it's given for public sector projects believed to be a waste of government funds. She also received a letter from a physical therapist in Colorado that said her time would be better spent running around the high school track herself. After all, breast movement seemed pretty straightforward—breasts move up and down. End of story. Why study it? "That's like saying there's impact on the feet when we run. We all know that, so then why should we research good footwear design and develop new cushioning materials?" Lawson told me. While breasts do bounce, researchers didn't know how to categorize that motion—in what direction, with what velocity, and with what force? "How do we understand what's going on with the breasts so we can counteract that [movement] through our [bra] design?" Lawson asks.

It turns out, it's really hard to study breasts in motion. They move in ways that are anything but straightforward. Not only do they travel vertically, but they also shift from side to side and move in and out, tracing a rough figure-eight pattern. On a graph, the movement looks something like butterfly wings. Breasts don't keep a uniform shape when they move through space either. Instead, they stretch and distort like a water balloon that's bounced up and down like a yo-yo. And no two breasts are alike, even in a given size. Tissue composition (fatty versus connective versus glandular), size, and shape differ tremendously among people.

It's this endless variation in chest topology that makes breasts a challenging area to study within the regimented field of biomechanics, hamstringing the field's progress. Normally, when researchers measure something like the angle of a knee, the variation between participants is low, generally around 1 to 2 percent. When it comes to breast motion, researchers can see variation closer to 10 percent between participants. It takes a lot of complicated math and protocols to accurately model and describe the trajectory of soft tissue that moves in complex patterns.

Researchers didn't always have the right technology to study breast movement in three dimensions. When Lawson started investigating breast biomechanics in the 1980s, she recruited fifty-nine women to run on a treadmill, both with and without a sports bra, and captured the sessions with a 16-millimeter camera. Every frame had to be hand digitized, and at one hundred frames per second, it took forever to process and analyze the thousands of frames she needed to see just one complete rise and fall of the breast. While it was a cumbersome undertaking, it allowed Lawson to begin understanding the vertical path of breasts.

But that's only half of the story. On film, Lawson saw that breasts moved laterally too, but at the time she didn't have the tools to quantify that motion. It turns out that vertical movement accounts for approximately 50 percent of breast motion during running, while side-to-side and in-and-out movement each account for 25 percent. Studies that look only at up-and-down displacement miss the bigger picture, limiting how bra designers and scientists understand the complex movement pattern and what it takes to create a supportive undergarment.

Even in the late 2000s, researchers like Jenny Burbage still came up against the constraints of technology when trying to capture all three dimensions of breast motion. Normally, researchers would place markers in relation to specific landmarks on the body and use a camera system to track the way those markers move during activity, both in space and in relation to other body parts. But sports bras by design cover the upper body landmarks researchers need to track. The motion capture system in Burbage's lab at the University of Portsmouth required that the markers be visible at all times, so they had to be placed on the outside of the bra.

That meant they weren't measuring how breast tissue responded and moved inside the sports bra, especially if the bra had molded cups. For example, if a medium sports bra fits a 36B and 36C, the 36C might fill up the cup while the 36B has extra room, a sensation that Lawson likens to a "bird trapped inside of a paper bag, flapping its wings trying to get out." That means the movement of the breast tissue can be very different between the two sizes. Other tracking systems involved heavy sensors that dragged down the breast tissue and influenced the breast's natural motion.

These methodological challenges may have also been a sticking point for scientific journals. Traditional biomechanics research methods are designed to evaluate rigid motion of the limbs and joints, but even by the mid-aughts there wasn't a consensus on the right approach to measure the movement of something like breast tissue. Until then, there were only a handful of scientific papers on the topic. "We kept getting rejected by reviewers," Burbage told me, and she suspects their methodology may have been the reason.

In order to investigate the questions they were interested in, researchers like Burbage and Brogan Jones, who runs the bra testing lab at the University of Portsmouth, have had to wait for technology to catch up. It wasn't until the mid-2010s that their lab got a new three-dimensional motion capture system where the camera didn't need to "see" the sensors to follow their movement. Instead, the sensors are tracked in an electromagnetic field. They're also small and light, allowing researchers to place an unlimited number on the body and under the bra. It significantly changed their ability to piece together the elaborate picture of how breasts shift in multiple planes during physical activity, and what's needed to support them. "We're progressing as the technology is progressing," Burbage says.

Lawson has come a long way since her 16-millimeter camera days too. She now uses an optoelectronic data collection system, the same technology used to make 3D movies and animation. With four cameras positioned in a semicircle in front of a treadmill, the cameras shoot infrared light at reflective markers placed on the runner's shoulders, collarbone, and nipples (on the exterior of the sports bra). The markers reflect light back to the camera, which collects three-dimensional coordinates fifty times per second. It doesn't take a picture per se, but Lawson sees the markers move on her computer screen. She has custom software that converts the data to an Excel file and extracts the rise and fall of the breast relative to the body. While Lawson doesn't put reflectors inside the bra, she does ask participants if there's excess room inside it.

Even as new technology has come into the field, Lawson sticks to a specific protocol she's used since the late 1980s to ensure she's comparing apples to apples across her studies. She even brings in one of the old Champion bras she tested in early studies to benchmark the updated system.

As researchers and designers better understand breast biomechanics, the style and fit of sports bras has improved. Champion Jogbra was the first company to commit to regular product research and testing. Shortly after LaJean Lawson finished her graduate degree and began her PhD in 1987, Lisa Lindahl and Hinda Miller from Jogbra called and asked if she could test their collection. Lindahl and Miller realized that different size breasts had different support requirements, and they wanted to understand how each bra worked for different people.

Lawson's research led to the development of Jogbra's motion control requirements, which are still in use. Based on a certain range of vertical displacement, bras are classified as high, medium, or low support. They even created a point-of-sales tool for retailers. It's a wheel where you select your size in one window and your sport in another window; the wheel then suggests the Jogbra style that matches your requirements.

It wasn't a given that brands would invest in rigorous scientific research on sports bras like they do with gear like sneakers. Sally Bergesen told me that when she founded the women's running apparel company Oiselle in 2007, there was no incentive for brands, especially major athletic companies, to focus on clothing, let alone sports bras. The margins were better with shoes, leaving apparel as an afterthought and women's apparel as an after-afterthought. What's more, Bergesen says bigger brands focused on novel styles instead of iterative design and innovation. "That's the kind of long-term thinking that a lot of gear makers miss because they're just so focused on the near-term results," she says. Sports bras still sold, even though they weren't great, because there was nothing else available.

Things started to change in the early 2010s, when workout clothes like leggings, tank tops, sneakers, and sports bras became the de facto uniform both inside the gym and yoga studio *and* out on the streets. Women-focused brands like Lululemon and Athleta became increasingly popular, and traditional athletic brands could no longer ignore the lucrative market. By 2018, the global athleisure market was estimated to be $300 billion; it's expected to reach $517.5 billion by 2025.

In the last five years, more brands have started to build their sports bras based on science. They're working with biomechanics labs at the University of Portsmouth and Progressive Sports Technologies in the United Kingdom, as well as consultants like Lawson, to test and validate bra designs, quantify motion control and comfort, and, most important, gather customer feedback on how the garments actually feel and perform on the body. Companies like Brooks, Lululemon, Nike, and Reebok have debuted new bras designed to perform and fit a wider variety of body types.

But science isn't always foolproof. In January 2012, women's athletic apparel company Moving Comfort (now part of Brooks) debuted its first biomechanically designed sports bra, the Jubralee. The end result was a super-supportive sports bra, one that locked and loaded the breasts in place and minimized movement. But it was built almost too scientifically—while it worked well once a person got into the bra, it wasn't comfortable for everyone, nor was it easy to find the right fit for every body.

Sports bra design and development remain a mix of science, art, and a touch of magic. With the availability of more research, the industry is just starting to understand body sizing, support needs, and the relationship between the two. Bra designers are still trying to find the sweet spot between comfort and support, a dichotomy that has vexed the industry and people with breasts for years. It's a tricky, inverse relationship: as support increases, comfort tends to go down. You could have a high-performing sports bra, but it wouldn't be comfortable. Or you could have a comfortable sports bra, but it wouldn't perform as well.

In the past, designers often felt the need to sacrifice comfort in service of support. "You can't just strap breasts down and that's it. What you're really trying to do is control that motion so there's less force, less of that start-stop impact," Laura Madden, director for apparel and bra development at Brooks, told me. She says learning that motion control doesn't equate to comfort, and that not all breast movement is bad, was a huge revelation. It's shifted how the industry thinks about sports bras.

For Lululemon, that means everything goes back to what the company calls "the science of feel." Lululemon's Chantelle Murnaghan told me that the intent of their bra design process isn't centered solely on achieving a specific percent of bounce reduction, but rather creating a certain feeling and experience for their customers. With the Enlite bra, which debuted in 2017, the design team tried to minimize the motion that wearers found uncomfortable while allowing for the motion that they did find comfortable. While the resulting bra's neoprene-like material looks hefty, it's soft and molds to the body.

Other companies like Brooks have attempted to reinvent the bra design process. In 2015, they decided to examine the assumptions that have underlaid bra design for the last one hundred years—what comfort means, what support means, and what each design element is supposed to do. Customers and their preferences have evolved. The current generation expects something different from their support garments than those before them.

The Brooks team started by going back to the basics. Helen Kenworthy, manager for bra development at Brooks, told me they wanted to know if they were overbuilding their bras. They cut up their products to understand the function of each design element and what it was really doing (or not doing). They tested bra linings to figure out what happened inside the garment and if they really needed two layers of fabric. They tested the bra cups. They reconsidered the use of seams, bindings, and finishes. They tested their existing bras for support and comfort in collaboration with the University of Portsmouth. Eventually, they came away with a scientific framework that set the rules of the road to guide their bra design going forward. Then they built the bras back up.

Previously, Brooks viewed bras as individual products, the fabrics, trims, construction, and cups designed specifically for each style. But given the diversity in people's bodies, breasts, and breast support needs, the company knew they couldn't just release one new style. They had to create an array of new bras. The result was the Dare run bra collection, which debuted in 2020, a suite of six bras—a scoopback, racerback, crossback, zip-front, high-neck, and strappy model—in sizes ranging from 30A to 44F. They don't look like your old compression sports bras;

they're encapsulation-style bras where molded cups separate and hold each breast individually, which studies suggest do a better job reducing breast movement than compression-based styles alone.

In creating the collection, Kenworthy told me they also let go of trying to achieve the maximum amount of bounce reduction. They instead focused on figuring out the amount of bounce reduction that equated with the best level of comfort, which also directed whether they got rid of or scaled down on components like bindings, trims, and finishes that could chafe. In 2021, Brooks launched a second collection of bras, the Drive collection, maintaining the same commitment. These four compression-based styles offer alternatives for people who don't love the encapsulation style.

While established brands began to rethink the science of sports bra design, they weren't moving fast enough for some people—particularly those with larger busts. Elyse Kaye first wrote a business plan for a sports bra company focused on curvy, larger-breasted women in the late 1990s. As a recent college graduate, she wasn't entirely serious about the idea and she figured someone else would create a suitable product before her.

But more than a decade later, when she trained for her first half-marathon, there were still no real options for her 32GGs. In the United States, people wear a DD cup size on average, but the industry's standard base size is a 36C. "None of the major brands were addressing it," she says. The problem is that though sizes can get bigger, they don't scale according to a neat mathematical formula. Grading sports bras for extended sizes requires skill, expertise, and a ton of back-and-forth testing. In other words, it calls for money and resources that companies aren't willing to invest.

The problem kept Kaye up at night. When she saw a man cocooning in an aerial yoga silk hanging from a carabiner, she realized she wasn't designing a bra in the traditional sense. She was designing a new kind of weight distribution product, one that would work for as many different body types as possible. She envisioned a bra with no underwire, one that pulled the weight from the front of the chest and off the bottom band and redistributed it to the back. In doing so, it would alleviate back pain, improve posture, and eliminate the dreaded red marks on the shoulders. She started thinking about pulley systems and shapewear, how each provided lift and allocated weight. Kaye enlisted the help of a corsetry expert, others from the shipping and packaging industry, and even someone from NASA. When she posted on a local Facebook group asking if anyone was interested in trying the bra, 165 women showed up at her house.

Kaye founded Bloom Bras in 2016, but finding factories to mass produce her product nearly proved to be impossible. Bloom Bras were a complicated design and ran in extended sizes—most factories preferred large orders that stuck to conventional sizes. After a factory fell through at the last minute, Kaye's production timeline was set back, and they started shipping product in 2018. When Kaye received her first shipment, she cried.

The garment is a front-zip bra that's easy to get on and off. It has wide, adjustable straps and a cincher in each cup so you can customize the fit to your individual breast. Seams that ride along the edges of the back are intended to carry the redistributed weight of the breasts. They currently offer sizes 28C to 56L, and Kaye says she can't keep the larger sizes in stock.

After more than forty years in the industry, LaJean Lawson still tests sports bras for Champion and says the garment has come a long way during that time. "It was really a new category that had to evolve in order for women to participate [in sports] with any decent amount of comfort," she says. She's excited about what's still to come, and that people aren't destined to be stuck with subpar sports bras. If you hate your sports bra, there's more information available to help you find a product you feel good in. "There just needs to be more dialogue," she says.

And with more research and technology, the industry has started thinking about what's next for athletic undergarments. Designers are just starting to use technology like lamination to make bras more body friendly. Some companies are pushing toward better moisture management and second-skin fabrics. Others, like Reebok, have incorporated smart fabrics into their designs. Called Motion Sense technology, the fabric adapts as needed. It firms up when it senses a lot of breast movement, say when you do box jumps or run, and then relaxes with less breast motion like during stretch sessions.

For elite athletes, bespoke bras may soon become part of their standard-issue kit. While most sports bra testing is conducted on a treadmill—running is an activity that's easily controlled in a laboratory setting and is a component of many sports—studies suggest that a good running bra might not be the best for all activities. Many run bras support and control movement in all directions equally. Women may need a bra that controls vertical support more when jumping and a bra that provides more side-to-side stabilization when doing agility drills.

Team GB (Great Britain) is betting on the untapped potential of sports bras to enhance performance. After a survey of seventy elite British athletes revealed that three-quarters had never been fitted for a sports bra and roughly one-quarter said breast discomfort inhibited them during training or competition, the English Institute of Sport (EIS) decided it was time to do something. They outfitted more than one hundred elite athletes, including Olympians and Paralympians, across fifteen different sports with custom bras as part of their effort to improve the health and performance of national team members.

Beginning in 2019, they worked with the University of Portsmouth and bra manufacturer Clover to determine the specific support and design requirements for different sports and provide breast health education to athletes. Field hockey player Sarah Robertson's new bra featured pockets to hold tracking devices like a GPS and a heart rate monitor, which she otherwise would need to house in a separate harness under her clothes. The encapsulated design not only supported her breast tissue but distributed it slightly forward to give her arms more room to move. Paralympic shooter Lorraine Lambert didn't need her bra to keep her breasts from bouncing; she needed it to keep her breasts out of the way of her rifle. She told *The Guardian*, "The gun is allowed to touch my hands because I'm holding it, and I'm allowed to support my arms on my torso, but the gun physically cannot touch me anywhere on my body." It if does, she could be penalized. Her bra design strategically positioned her breast tissue so that it wouldn't accidentally brush against the rifle, which would also affect her shooting accuracy too.

The recognition of sports bras as essential gear is long overdue. People with breasts just want something that works for their bodies, and brands need to commit to making good-quality, affordable bras that fit a diverse range of shapes and sizes, particularly people with larger busts. The right sports bra could be a game changer. It would break down a major barrier to fitness and sports and could unlock another level of confidence. Lawson says, "Every woman deserves the considerable physical, mental, and emotional benefits of exercise, and she needs to have a sports bra that will let her do that."

Sports bras changed the relationship between women and physical activity, making it possible for women to move more freely, with less pain and embarrassing motion. What started with a "frankenbra" sparked a conversation around what it means to design athletic products specifically with women in mind. Now this revolution is spreading to other areas of the outdoor and sporting goods industries.

8.

Beyond Shrink It and Pink It

• • • ● ● ● • • •

C harlotte Dod's nickname was "Little Wonder," a nod to her superstar athletic accomplishments during the Victorian era. She was an Olympic medalist in archery, a national-level field hockey player, expert mountaineer, and adept figure skater. But she was best known as a tennis player, having claimed the first of five Wimbledon titles at age fifteen. She won her matches on the grass courts of the All England Croquet and Lawn Tennis Club while wearing long, high-necked dresses, thick stockings, leather shoes, and a corset. She wasn't entirely thrilled with her attire, once telling a journalist, "How can [women] ever hope to play a sound game when their dresses impede the free movement of every limb?"

While athletes no longer have to wear long skirts and corsets, frustrations with clothing and gear still plague women. The sporting goods and outdoor industries continue to be dominated by men—from the design, merchandising, and marketing teams to the executive suite to the "ideal" consumer, the vast majority of athletic gear is created by and for men. For decades, women have made do with men's and unisex gear, getting used to settling for good enough as long as it was functional.

That often meant boxy ski parkas, hiking pants that gap and bunch in funny places, and backpacks that are too loose even with the straps cinched all the way down. Cold-water surf gear, like booties and mittens, can be a little too big or wide, making it awkward to grip a surfboard and harder to stay warm. Even steering pedals on one-person outrigger canoes can be out of reach, leaving some women to rely on foam blocks if they want their feet to actually reach the pedals.

A few companies did cater to the needs of women. Jogbra and Moving Comfort began selling sports bras and running shorts, respectively, to active women in 1977. Fitness culture boomed in the 1980s, riding the coattails of Olivia Newton-John's music video "Physical" and Jane Fonda's workout videos. A whole market segment of shoes and clothing emerged to outfit the women who flocked to classes like aerobics and Jazzercise. Reebok launched its iconic Freestyle sneaker in 1982, and by 1984 the soft leather aerobics shoe made up more than half of the company's sales. (Actress Cybill Shepherd even wore a pair of bright orange Reeboks under her gown at the 1985 Emmy Awards.)

Since many of these exercise programs drew from traditionally feminine spheres like dance, leotards and leg warmers were staples in class, where the emphasis was on sculpting a sexy, trim body, not turning women into professional athletes. Even sports bras, which started as purely functional, became more of a fashion item, complete with a plunging neckline. Because of the form-fitting nature of activewear, fitness historian Natalia Mehlman Petrzela says, "you see them being marketed, designed, and sold in a way that isn't necessarily about prioritizing women's athletic function but more about having them look pretty."

In the 1990s, group fitness remained popular, but women athletes began to rise to the forefront. Nike launched its "If You Let Me Play" campaign, touting the power of sports to transform young girls into powerful, confident, and resilient women. Then, in 1999, the U.S. women's national soccer team won the World Cup on home turf, and everything changed. "There was a real understanding of what female athletes could do and how important it was for young girls to look up to female superstar athletes," says Lucy Danziger, former editor in chief of *Women's Sports and Fitness* and *Self* magazines.

Athletes like Mia Hamm, Julie Foudy, and Brandi Chastain became household names. Brands realized that women were a potent, emerging market segment in the sporting goods industry and rushed to adapt their products. You can't blame companies for wanting to tap into the women's market, which Missy Park, founder of women's activewear brand Title Nine, describes as "a big damn market." She says, "Money talks and people are paying attention to that."

Park knew that women were an overlooked and underserved segment in the industry. She was part of the first cohort of women to go through high school and college with Title IX in place. While it meant she had opportunities to play organized sports—and she played all of them, from lacrosse to tennis to basketball—it didn't necessarily mean that schools and colleges were ready to accommodate women athletes. As members of the women's basketball team at Yale University, Park and her teammates were stuck with hand-me-down uniforms from the men's team. Players with smaller-size feet wore kids' basketball shoes because there were no women's sneakers. "We had to scratch and claw pretty hard," she told me. "Every win we got in terms of practice time or facilities or equipment felt like a loss for the men's team." That experience, along with her post-collegiate career working in the outdoor industry for companies like the North Face and Gary Fisher Mountain Bikes, planted a seed in Park's mind. Women needed gear of their own—the "women's version of Nike."

As Park tells the story, starting Title Nine in 1989 was more of a selfish act. She just wanted a place to buy the clothing and gear she needed for her sports and outdoor adventures. At the time, there was no Lululemon, Athleta, or Sweaty Betty. Back in the late 1980s, retailers weren't exactly clamoring to serve women and girls because they weren't streaming into stores demanding performance-oriented gear. But manufacturers, who forecast several years into the future, did see the massive hole in the market and the growing demand for women's products on the horizon. "I was able to jump in and take advantage of the fact that manufacturers were making stuff that most retailers weren't willing to buy," Park says.

Park's hunch paid off. She launched Title Nine as a mail-order catalog in 1989, sending out thirty thousand catalogs filled with running shorts, tights, and sports bras. Four years later, Title Nine turned a profit. Today, they are the largest woman-owned-and-operated retailer in the United States and the women's sportswear space, a market estimated at $26.8 billion in 2018.

Though the industry began to recognize that women offered an untapped market opportunity, the standard approach was to take gear originally designed for men, make it smaller, add a cap sleeve or nipped waist, and offer it in a "feminine" color or a floral pattern. It's a strategy referred to as "shrink it and pink it," and it often fell flat. These derivative products treated women's needs and lived experiences as an addendum.

"It was painfully obvious that the sporting goods industry wasn't ready or prepared for women athletes, especially athletes that expected a high degree of technical performance," Missy Park says. "What keeps women comfortable in the outdoors is very different than what keeps men comfortable in the outdoors." Women's bodies are not anatomically and physiologically the same as men's, from breasts to (generally) wider hips to different metabolic rates, which can influence body temperature. Women need gender-specific apparel and gear to account for these differences. "Women wanted it all: fit, comfort, performance, and style.

The companies that got it right are the ones still in business," says Lucy Danziger.

Even as more gear trickled into stores and online shops, women and girls still weren't valued as equally as men, leaving them with a painfully limited selection. When Heather Malloy started playing recreational hockey in 2006, the New Jersey resident couldn't find protective equipment that fit her. Most pads were sized for a man, including those labeled as unisex, and they didn't work for someone with breasts. While women's pants and shoulder pads existed, there were usually only one or two options, while there were dozens of styles and sizes for men to choose from.

"I'm a smaller-framed person and I don't need to be kitted out like a Transformer to play beer league hockey," Malloy says. She couldn't maneuver well, often feeling like a "toddler in a snowsuit." She has worn a combination of men's and junior's equipment over the years, altering her shin guards and elbow pads to get rid of some of the bulk. Kennedy Birley faced a similar challenge when shopping for soccer goalie gloves as a kid living outside Portland, Oregon, in the late aughts. Birley wore boys' gloves because there was a wider range of sizes and better-quality options. But the gloves were too big. It changed how Birley defended goal, leaning more on deflecting the ball rather than catching it. "Gloves that were too big meant that I didn't trust the ends of my fingers because I didn't know where they were in the glove," Birley says. "I would definitely get more anxious and was more uncomfortable when my gloves didn't fit."

In some cases, there weren't—and still aren't—choices at all. Shockingly, most major brands don't offer a women's-specific soccer cleat, despite soccer's worldwide popularity and FIFA's goal to double the number of women players to sixty million by 2026. Instead, women wear men's or kid's shoes. "You walk into the sports store and it's like men's, kid's, and that's it. So, if you're a female, do you even exist in the sport?" says Laura Youngson, co-founder of the shoe start-up Ida Sports. What surprised Youngson was that the lack of gender-specific shoes wasn't just a problem in soccer. Folks from a variety of sports—mostly those traditionally played by men like basketball, cricket, rugby, baseball, and lacrosse—reached out and asked her to design a shoe for them.

It wasn't until 2019, in advance of the Women's World Cup, that Nike unveiled soccer uniforms specifically designed for women, not just a spin-off of the men's kit. The jersey has a longer sleeve to minimize exposure of the upper arm. The neckline is a cross between a crew and a V-neck so athletes can easily pull it over a ponytail. U.S. Women's National Team player Megan Rapinoe said, "The more I'm messing with [my uniform], the less focused I am, so if I put something on and don't have to mess with it, I'm just locked in and ready to play."

The demand for women's-specific products clearly isn't just a matter of people being picky or difficult. The lack of appropriate gear and apparel can hamstring one's sporting ambitions, confidence, and performance. Clothing and gear also need to accommodate diverse body types across the gender spectrum, not just the narrow range of bodies historically seen in marketing campaigns and ads. When people can't find gear that suits their bodies, needs, or identity, it becomes a barrier to entry, effectively erasing their place in sports and the outdoors. When the writer Latria Graham showed up for a cover shoot for an *Outdoor Retailer* magazine feature on plus-size clothing, only one outfit out of one hundred pieces of clothing fit her, despite asking for extended sizes. The implications of these limited choices are real. Graham writes in a post on Instagram, "It's driving 6-hours round trip (pre-pandemic) hoping that one of the two pairs of pants in stock will work for the technical hike I'm doing on a tight-turnaround assignment that weekend . . . The more dangerous the activity—diving or mountaineering—the riskier my clothing choices become. Pieces that are too long are hazards. But they're all I have. I need to be in that water or on that mountain to do parts of my job and I take those chances."

Without an adequate selection of sizes, shapes, and technical specifications to choose from, women end up spending a disproportionate amount of time searching for options and rigging their equipment so it fits their body the way they need it to. The consequences are, at best, a lack of aesthetic appeal and, at worst, discomfort, pain, and injury that can affect girls' and women's performance or even their willingness to participate in sports.

In the last decade plus, the tide has started to turn thanks to the emergence of new women-owned companies and a commitment among brands and retailers to raise the profile—and quality—of products made for women. (It doesn't hurt that by 2028, women are projected to account for 75 percent of overall discretionary spending.) For example, in 2017, REI launched its "Force of Nature" campaign, retooling products like sleeping bags and hiking backpacks to better suit women's bodies and expanding its apparel sizes, particularly in technical categories like rain and insulated jackets, snow pants and bibs, and hiking clothes. At the same time, others are thinking more deeply about what gender-specific distinctions really mean and when they are warranted and necessary. "Many companies see that they're missing the mark massively if they don't listen properly, actively, and strategically to women," says Edita Hadravska, former design director at Arc'teryx.

There are many reasons why the "shrink it and pink it" strategy made practical sense. Companies had established manufacturing and supply chain pathways. They had patterns, fit models, and molds for apparel, shoes, and gear that already worked for their men customers. Surely those general dimensions and geometries would work for women too. Why reinvent the wheel? Doing so would be expensive, and there was no real incentive to revamp the existing business model or research and development process.

But in the process of grading down clothing, accessories, and shoes to a smaller scale, certain features hit the cutting room floor because there's not as much material to work with. Outdoor gear tester Eve O'Neill told *Medium* in 2018, "[I] can't tell you how many times the women's versions of things I've tested have been missing features that are on the men's versions," things like functional pockets or belt loops or zippers. Some women's hiking boots are rated worse compared to men's versions because there's less surface area for the lugs needed to gain traction in mud and dirt.

The shrinking process can also lead products to be misaligned with a woman's form. Take hydration vests: with a narrower cut and smaller proportions, pockets can end up positioned on top of or to the side of the wearer's breasts, which adds weight and bulk right where most people don't want it. Other times, products ignore physiological sex differences entirely. When it comes to regulating body temperature, women tend to feel colder than men—yet product testing for their clothing is done from a man's perspective. "Even up to today, for example, typical heat lab manikin testing is done on male manikins," says Hadravska. "You end up having to test sized-up garments and deal with other 'noise,' which wouldn't be there if you had a manikin with proper women's proportions."

When women's gear is put through a male-centric product development process, it doesn't fully consider what women want and need from their apparel and equipment. Instead, men's priorities and gendered assumptions about women's bodies continue to steer the design conversation. It creates a glass ceiling that can hold women back from their full potential.

The bike saddle, for instance, was designed to suit the men who historically dominated the sport. In the late 1990s, these cyclists complained of numbness and pain while riding, even erectile dysfunction. The revelation led physician and ergonomic designer Roger Minkow to create the first anatomically based saddle. It had a cutout along the center to eliminate the excess pressure and increase blood flow to the genital area. When companies decided to make a saddle for women, they adapted the existing model by enlarging the size of the cutout to accommodate women's genitalia.

The problem is, women's anatomy isn't the same as men's, which is why pro cyclist Alison Tetrick could never find a comfortable seat. Having leapfrogged from amateur triathlete in 2007 to USA Cycling National Team member in 2008, Tetrick didn't have the trial-and-error experience most recreational and up-and-coming riders had testing gear and figuring out what worked for her body. Instead, she relied on the advice of experts around her. She kept telling bike fitters that her saddle hurt, but they responded that it was normal to be uncomfortable. In a sport where suffering is part of the ethos, she accepted that cycling would be far from a cushy ride.

With the right fit on the saddle, the body will naturally relax into its optimal riding position—which affects efficiency, power, and performance. Tetrick's saddle discomfort, however, prevented her from finding the best position on her bike. To ride aerodynamically, she had to roll her pelvis forward onto the nose of the saddle, which she describes as a "tiny, narrow thing with no padding." For women cyclists, that position scrunches up the genitalia and places intense strain on the soft tissue of the labia, tissue that isn't meant to bear weight. To relieve the pressure, Tetrick sat crooked, rolling her pelvis back and tilting it to the left.

These small accommodations have big biomechanical implications. According to Andy Pruitt, founder and retired director of the Boulder Center for Sports Medicine and a bike fit expert, changing the position of the pelvis and lumbar spine disengages the glutes, the powerhouse muscles critical for generating power on the bike. "Here are these women trying to get low in the front but reversing their pelvic posture to relieve the saddle pain. It's a huge battle going on," he told me. "If you can't sit in the saddle comfortably and evenly, it goes everywhere. It can be your back. It can be the lower extremities." Tetrick believes her torqued pelvis and compromised form led to pain in her knee, IT band, and back.

To make matters worse, Tetrick noticed swelling in her genital area that quickly ballooned. While the initial throbbing subsided, there was a constant level of inflammation and trauma to her vulva, the external female genitals, that worsened over the years. Tetrick says she wasn't the only cyclist who encountered these issues—some women also experienced numbness and damaged lymph nodes and vessels. "We would talk internally about the saddles ruining our sex life or causing swelling and disfigurement," Tetrick says. Post-race, riders would don multiple pairs of Spanx, stuff ice down their shorts, and put their feet up in search of relief.

Recreational riders can hop off the bike after an hour and take a few days or a week off to let the tissues around the vulva recover. Elite athletes, however, can't stop riding. The problem gets progressively worse until cyclists need labiaplasty, a procedure where a plastic surgeon trims excess skin from the labia. After seven years as a pro cyclist, Tetrick opted for surgery. Soon, she learned that a number of other pro riders had chosen to get the same procedure, sometimes multiple times. It left her wondering: Why aren't we talking about this? Why isn't anyone doing anything about it? It's the gear that should be fixed, not vulvas.

Similarly, people with breasts who participate in sports like martial arts, fencing, and hockey need chest protection designed to accommodate the anatomy and contours of their torsos. Retrofitting men's chest protectors doesn't really work. While we may joke that getting knocked in the boobs isn't a big deal because it's just soft tissue, breasts are vulnerable to bruising, swelling, trauma, and even fat necrosis because of their extensive network of small blood vessels. And those direct blows hurt. In a study of elite Australian athletes, 36 percent reported sustaining breast injuries and 21 percent said it affected their performance.

Likewise, a person's body continually changes during pregnancy and after childbirth. While pregnant and postpartum people are encouraged to exercise to reap a multitude of health benefits, activewear options are sorely lacking. Workout pants dig into the waist and legs. Sports bras feel even more constricting than usual, given pregnant people's expanding rib cages and extra-sensitive breasts, and don't offer easy access for nursing parents. The material isn't great either, leaving many overheated because it doesn't dissipate sweat. Without good choices, people are left to either exercise in subpar apparel or not exercise at all because they're so uncomfortable. Major brands only recently launched maternity collections—Reebok in 2019, Adidas in 2020, and Nike in 2021—while other companies including Athleta, Lululemon, and Under Armour currently don't carry items for those who are expecting.

While saddles, breast protection, and maternity apparel are clear-cut areas where women need gear tailored to their anatomy, a less obvious example is shoes. Women's feet aren't just smaller versions of men's feet. In general, women's feet are V-shaped, with a narrower heel compared to the forefoot. But men tend to have longer, more rectangular-shaped feet. Women also have a higher instep and different spacing between the midfoot bones. The problem is a shoe's last—the mold that shoes are built around—is often based on the proportions of men's feet. That goes for so-called unisex lasts too. "It's not like they said, 'Oh, let's split the difference between the genders.' It's just male and then they say it's going to fit a man and a woman," Geoffrey Gray, founder and director of research at Heeluxe, an independent footwear testing company, told me.

As a result, many athletic shoes don't fit women's feet. "We've tested the same shoe on men and women runners, and we've found in our research it's eighty percent tighter on the women's forefoot compared to the men in the exact same shoe," says Allison Meadows, head of data science at Heeluxe. Meadows says women assume that their feet are weird or too wide when in fact their feet are just their feet and a normal width for a woman.

When I asked Gray why the footwear industry doesn't routinely build shoes on women's lasts, he told me that athletic footwear is "an old boys' club" and read me a quote from Geoff Hollister's book, *Out of Nowhere: The Inside Story of How Nike Marketed the Culture of Running,* that sums up the sentiment. Hollister was one of Nike's original employees and wrote, "The Lady Cortez was one of the first athletic shoes made on a women's last, and its success showed just how big that market was. Being the dumb bunch of guys we were, we didn't get the message right away." Gray says executives assumed they knew all they needed to know about making shoes and didn't listen to women. On the manufacturing side, it's easier to keep everything uniform and use the same last. "Making these small but significant changes to footwear to create similar experiences for women and men is tougher for the factory," he says, not to mention more expensive.

Today, more shoes are marketed as woman-specific, particularly among running brands. But it's not always clear what that means. When Brian Beckstead co-founded footwear company Altra in 2009, he told me that hardly any brands talked about what gender-specific meant in terms of footwear or did anything about it. Beckstead says there are three main tiers to making a women's-specific shoe. Aside from the last, there's the shoe's midsole. If companies use a gender-specific last, the molds and tooling also need to be customized for each gender. There's also cushioning—since men tend to be heavier than women, the midsole foam is denser and firmer in men's shoes and softer in women's shoes.

Altra has designed shoes with women's- and men's-specific fits since its beginning and invested in the midsole and outsole construction needed to support the different fits. Other brands might just tweak one or two features—like adjusting the cushioning, a cheap and easy fix—and call their shoe gender-specific. But now, Beckstead sees more brands making fit adjustments based on women's feet. He says, "We need to continue to put pressure on other brands to continue that progress of making products more gender-specific."

Cyclist Alison Tetrick's story is an extreme example of what happens when women are shoehorned into a male-centric design model. While brands are moving away from "shrink it and pink it" and embracing the concept of women's-specific design, it's still far from perfect. The industry continues to employ a deficit-based approach, where women are positioned as lesser than men. "Men are still this pillar of normalcy and everything is compared against them," says Jen Gurecki, founder of Coalition Snow, which makes skis and snowboards for women.

That's in part why when women-specific skis first started appearing on the market in the mid-1980s, they were watered-down versions of men's equipment. They were shorter and softer, requiring less strength and weight to turn and control on the slopes, and Gurecki told me they "sucked." Implicit in the design was the assumption that women weren't as strong skiers as men. They didn't need the longer, stiffer planks that men were accustomed to and were better suited for aggressive skiing and varied terrain. "You get sandbagged before you even get out the door," Gurecki says.

Casey Kerrigan, physical medicine and rehabilitation physician and founder of OESH Shoes, sees a similar dynamic at play in footwear, an assumption that differences between men and women somehow mean women are flawed and need to be fixed. As an example, she points to women's Q angle, the angle formed between the quadriceps muscle and the tendon in the kneecap. On average, a woman's Q angle will be higher than a man's because women have wider hips. It's been hypothesized that a higher Q angle contributes to knee injuries due to greater stress in the knee and increased pronation (when the foot rolls in upon landing). Shoe companies then market shoes to women with side-to-side contouring and guide rails to control foot pronation and "correct" the problem. But when Kerrigan studied the effect of running shoes with these stabilizing features on stress in knees, the shoes actually changed the body's mechanics and movement pattern for the worse, in both women and men. Yet, when women and men walked barefoot, stress levels weren't elevated in either group. More recent studies have shown no association between higher Q angles and knee injuries in women. "Whatever women are doing, we're doing fine as long as you don't mess with it," she says.

When it comes to clothing, the traditional apparel design process doesn't take into account how bodies change, particularly as they go up in size. Most clothing is designed around an ideal fit model, a person who represents the supposed middle of the size range, usually a size six for women's clothing. Designers make a pattern, which is then mathematically sized up or down in a process called grading. They spread the pattern pieces apart to make bigger sizes and overlap them to make smaller sizes; the space between the pieces is governed by predetermined rules.

While the grading process works for sizes close to the fit model, things get wonky with plus sizes. As pattern pieces are spread farther and farther apart, "you're so far away from the original, and it no longer works," explained Raquel Vélez, founder of outdoor apparel brand Alpine Parrot. Sleeves end up too long, tops too boxy, pants too big at the waist, and necklines so wide they fall off the shoulders. "It's just a poor understanding of how bodies change as they change sizes," she says, because bodies don't scale proportionally. Unsurprisingly, when larger sizes don't sell, the athletic and outdoor industry assumes there's no market for these clothes.

But there is a market—a big one. The plus-size market is one of the fastest growing apparel segments in the United States and was valued at $24 billion in 2020. An estimated two-thirds of American women wear a size 14 or above, with the average size between a 16 and 18. Yet many brands haven't been willing to adjust their design process or find fit models who better represent people who wear larger sizes. In the summer of 2020, Vélez found that only 10 percent of hiking clothes on REI's website were available in extended sizes. A recent survey of skiers, snowboarders, and outdoor enthusiasts found that 97 percent of plus-size athletes had trouble finding winter outerwear that fits, and 60 percent said the lack of well-fitting gear kept them from participating in sports.

To Vélez, it sends the message that she and those like her should be grateful they have an option, even if that option may not fit. But it means she spends her time on the trail pulling up her pants because the waist doesn't fit or moving uncomfortably because the pants are too tight through the hips. She almost gave up skiing altogether when she couldn't find size 16 snow pants. The lack of options means there isn't equal access to the outdoors. "It's this idea that you're not allowed to engage in a thing that your friends engage in," she says.

Jenn Kriske is the founder of the bike apparel company Machines for Freedom. It doesn't surprise her at all that sporting goods companies have long treated women as a meddlesome exception to the rule, given that all-men design teams are de rigueur in the industry. It's like playing a game of telephone when designers themselves don't wear or use the products they're creating. Women product testers may not convey the full story because it can be embarrassing to discuss pain and discomfort with men, especially in the breast or genital region. Plus, people often struggle to put their experience into words, something that Kriske learned from her years in hotel and restaurant design. "As a designer designing for other people, so much of it is about the psychology of just getting to the heart of what they really mean," she told me. "They could say, 'This is a little bit uncomfortable,' but they can't really get into detail about why they like something or why they don't."

Most of Kriske's bike kit options were scaled-down versions of men's clothing and weren't designed for technical performance. She didn't love the way she felt in them. While her guy friends debated outfit choices from their extensive cycling wardrobes, she scrambled to find enough gear to get her through a few days of continuous riding. The breaking point came after a 120-mile training ride in the bike mecca of Marin County, California. A seemingly normal saddle sore grew to the size of a lime overnight, and Kriske understandably panicked. She couldn't cross or even close her legs, and the infection almost landed her in the hospital. It turned out that the chamois, the padded crotch section of her bike shorts—shorts made by a brand that her friends who were men touted as the best—was misaligned. To alleviate pressure on the crotch while in the saddle, the thickest part of the foam section is supposed to be aligned under the sit bones. If it's not, Kriske says, "You have bunching where you don't want bunching, and then you don't have support and padding where you need the support and padding," leading to saddle sores.

Kriske wanted to take the game of telephone out of the design process. She left her job to found Machines for Freedom in 2013 and create cycling bibs (essentially a pair of bike shorts with spandex suspenders attached) cut for a woman's shape. She worked with a bike fitter in the Los Angeles area and analyzed years' worth of data. What she learned was that, while saddles come in different sizes to fit a range of sit bone widths, most chamois didn't accommodate the widest sizes and left riders vulnerable to chafing and sores.

She ordered every pair of women's bib shorts on the market at the time. Most fit poorly. The chamois was too narrow or positioned too far forward or too far back. The fabric on some bibs would start to stretch and become transparent—not a great look when riding in an aerodynamic position for hours on end.

Kriske then built prototypes for her cycling bib. Not only was it paramount to get the chamois sizing, foam density, and materials right, but she also wanted to dial in the fit and compression, eliminating the dreaded "sausage legs." Most brands relied on super-stretchy fabrics that can fit over any body shape and hip size. To hold the short down, they place a tight elastic band at the bottom of the hem—and the fleshiest part of the leg. Kriske used a different approach to achieve allover compression through the hip and leg and eliminated the need for the rubber band around the thigh. "People called our stuff Spanx for the bike," she says, which can feel amazing when you're hunched over a bike for hours. The bib's bodice also accommodates a variety of heights and bust sizes.

Kriske stopped getting saddle sores wearing her Machines for Freedom kit, and she doesn't even need chamois cream anymore. "All of those problems just went away. It's like, Oh, I didn't have to suffer like this. No, this isn't part of the sport," she says. As soon as she launched the product, she received tremendous feedback. "The community was like, 'Oh my god, finally! Somebody finally gets it,'" she says.

On the saddle front, while the chronic inflammation women experienced was an open secret among elite cyclists, team directors and brands were clueless. Specialized, one of the biggest bike companies in the world, caught wind of Alison Tetrick's story and asked her to come in and discuss it. In 2016, the company enlisted Andy Pruitt, a frequent consultant to the company, to head up a project to design a new women's saddle.

Pruitt took a decidedly scientific and women's-specific approach that blew up the traditional saddle design. He had to start from scratch because standard instruments like pulse oximeters, neurometers, and pressure maps didn't accurately read the way female genital tissue reacted and responded to saddle pressure. Pruitt also turned to another field for inspiration—prosthetics. Having lost his lower right leg, he was intimately familiar with the limb-fitting process. "The residual limb is not meant to bear weight, and the crotch tissue is not meant to bear weight," he told me. When fitting someone, prosthetists use what's called a check socket, a clear plastic device that allows them to see how the prosthetic limb interacts with the skin tissue of the residual limb. It also shows if and where blood flow is restricted (ischemia) and where there's excess fluid (edema).

Pruitt adapted this idea to the saddle project. In the lab, a team of women volunteers rode naked on a stationary bike fitted with a clear plastic saddle. "I could see the ischemia under the sit bones, and we could see the labia becoming engulfed in fluid and actually swelling and descending into the cutout [of the saddle]," he says. It happened even when riders wore bike shorts. Tetrick participated in the saddle design project too. In one test, researchers used a putty-like material to determine how tissue around the genitalia made contact with the saddle and how it moved while riding at different effort levels. "It was so validating to see where my skin and soft tissue were going and why damage was happening," Tetrick says.

With this information, Pruitt crafted a new saddle. It incorporates what Specialized calls "MIMIC technology." In place of the cutout, there's a trampoline-like base at the center to accommodate the variety of shapes and sizes of riders' vulvas and gives the external genitalia a place to rest. Denser foam at the rear of the saddle supports the sit bones while softer foam pads the nose. Since riding with the new saddle, Tetrick's swelling hasn't returned, even as she's moved on from pro road cycling races to 200-plus-mile gravel events, earning her the title "the Queen of Gravel." When the saddle debuted in 2018, it sold out in about ten days.

The Specialized saddle and Machine for Freedom cycling shorts and bibs are two examples of solutions that would have never come about without considering women first—their strengths, anatomy, physiology, and needs. The research and design process was an entirely separate and distinct one that centered on women. "Once you can open the dialogue and make people feel heard or have a safe place to go, that's where innovation and development can occur," Tetrick says.

Laura Youngson created a similar space for soccer players to talk about cleats. She always wore kid-size shoes, made with "rubbish material." She was appalled to find out that even professional players were left with only men's and kid's models. She explained the dilemma players face: They'll either get shoes that fit the heel but leave their forefoot and toes squished and prone to black toenails. Or they'll choose shoes that fit the forefoot but leave the heel loose and susceptible to blisters. And in a dynamic sport like soccer, which requires players to accelerate to top speeds and change direction on a dime, ill-fitting boots are problematic. Women players will shave down cleats, particularly underneath the ball of the foot, to alleviate pressure points. Others use duct tape to make alterations to the shoe's fit. Athletes may not share their complaints or DIY solutions with the brands, for fear of being dropped from their contracts. "We finally get free boots. I'm not going to tell you all the things that are wrong with it," Youngson told me.

While Youngson acknowledged that there are limited scientific studies on the relationship between shoes and injury, particularly among women soccer players, anecdotal evidence from the pitch was enough for her to try to do something different. "We do know that comfort and fatigue are really closely linked and that fatigue links to injury," she says. "If you're comfortable in your boots over ninety minutes, you're less likely to be fatigued. If you're less likely to be fatigued, you're less likely to be at risk of injuring yourself." As an outsider to the industry, Youngson rewrote the rules of shoe design when she started Ida Sports. She believed that if she could make a more comfortable boot, it would be the first step toward injury reduction. Then she could help secure more funding for more research.

Youngson and her co-founder Ben Sandhu made the first prototypes in her kitchen, pouring resin into a mold. A year and a half later, in February 2020, Youngson and her team released their first soccer shoe. Ida's shoe has a wider toe box for better weight distribution across the front of the foot and a rearranged pattern of studs under the ball of the foot to account for the specific way women pressure load their feet quite differently from men. It has a narrower heel cup and bends at a slightly different place compared to unisex shoes to alleviate unnecessary pinching and discomfort. The shoes have allowed women and girls to play without foot pain or blisters, including players in Australia's A-League (the top women's soccer league) and AFLW (the top women's Australian rules football league). They no longer worry about their boots and can focus on the game.

Raquel Vélez wanted to address the lack of diversity in sizes in the outdoor apparel industry, starting with hiking pants. As a former mechanical and software engineer, Vélez had plenty of experience taking a minimally viable product, figuring out the users, and then "testing the shit out of it," she told me. She knew the apparel design system was broken, but she needed good data to create a new one. Using herself as a fit model, Vélez created a pattern and graded it for sizes 14 to 24. But she didn't trust traditional grading rules and recruited ten fit testers in the Bay Area in California, eventually learning two things. One, she needed new grading rules; and two, not everyone is shaped like her.

Vélez describes herself as having "a big booty and a smaller waist," but other women are more straight up and down with similar-sized waists and hips. She needed two different fit styles. She created another batch of pants with the new fit and grading rules, tested the pants with more people, and modified her pattern again.

Even wearing the imperfect prototypes, testers loved the pants. Vélez recalls hiking with one woman who didn't have any real complaints during their time together. After they parted ways, the woman texted Vélez. She said that for the first time in her life, she didn't think about her pants, whether they were chafing, or if she had to pull them up or down. She just *enjoyed* their hike. "That was the moment when I realized I'm not making just pants. I'm creating opportunities for joy," Vélez says. Others saw the clarity of her vision too—and the neglected market. She was a finalist in Title Nine's Pitchfest, a competition for women entrepreneurs, even though she didn't have any product. She launched a Kickstarter campaign in April 2021 with plans to go beyond size 24.

Vélez named her company Alpine Parrot after a species of parrots that live in the Southern Alps of New Zealand and make snowballs with their friends. While their exterior feathers are a drab olive-green color, their underwings are rainbow colored. When they fly, you see an explosion of color. "When I thought about my market, I thought about how often people of size and people of color have the stereotype that we don't go outside. But when we are in our element, we have this incredible joy," Vélez says. "This is the perfect analogy of an animal that defies expectations, defies stereotypes, and when it's in its element, it truly shines. That's us."

While there has been more investment in women's products, I've long wondered what we truly mean when we designate something as "made for women." Companies aren't always forthcoming with the research or data to back up their claims. Is it based on some physiological or anatomical need? Is it a marketing gimmick? Will the clothes fit better because a person has wider hips and breasts? Is the material different from the men's version? If so, why? And when does a gendered product actually make a difference?

Frankly, not everything needs a gender-specific approach. The choice of tents and camp stoves, for example, doesn't depend on one's anatomy or physiology. Nor do surfboards, where different volumes, lengths, and shapes suit different waves and style preferences.

While companies like Altra and Ida Sports are committed to a women's-specific design strategy, others are eliminating gendered distinctions when data and science warrant it. By decoupling gender from certain equipment and gear, people have more choices across the board. For instance, Specialized has gone all in on the MIMIC technology and new saddle design where physiology and anatomy require it and it makes a difference in terms of comfort, performance, and injury prevention—but they've pulled back on creating women's-specific bike frames.

Todd Carver, head of human performance for Specialized, explained to me that the general idea for women's-specific frames was based on the assumption that women had longer legs and shorter torsos compared to men. But there wasn't much data to support or refute that strategy. In his opinion, it was more of a marketing campaign to attract women. The idea didn't sit right with him as a former bike fitter. In 2017, he sifted through the anthropometric data from roughly eight thousand bike fits from Retül, a three-dimensional bike fitting platform that he founded in 2007. "What I found through my research into the Retül database is that the variability within a gender is much greater than the variability between a gender, between sexes," he says. In other words, the bell curves of men and women overlap far more than they don't—men and women have similarly proportioned bodies when it comes to leg and torso lengths.

In 2018, Specialized stopped making women's frames across their family of road, race, and mountain bikes based on the bike fit data and the feedback they received from riders and retailers. "We want to invest heavily in R and D where there is a difference and really make products for specific sexes. But where there is no difference, we are not going to make sex-specific product," he says. If you pull back the curtain and look at all the available bike geometries, every rider has more choice. Once a rider is matched with a frame for their size, they can personalize the fit and adjust specific touch points where a gender-specific approach matters—saddles, handlebars, grips, and cranks.

Similarly, some ski companies have started to ditch the women's and men's labels on their planks. Austrian sports equipment company Fischer's "Skis for Skiers" initiative wants to give all skiers the equipment they need to tackle the mountain, whether on moguls, blue squares, or backcountry. They use the same construction in both men's and women's models. They believe the choice of ski depends on a skier's height, weight, performance level, and discipline, whether touring, piste, racing, or freeride. Former Freeride World Tour competitor and editor at *Backcountry Magazine* Louise Lintilhac told *GearJunkie* in 2020, "It doesn't matter what gender you are, body type is body type. Your skiing ability isn't gender-specific. We should be marketing to body weight and ability, not gender." Jen Gurecki of Coalition Snow says the best ski for you depends on how you ski and your goals. If you're a powder hound, you need a fatter ski to float over the deep snow. If you prefer moguls, you need a ski with a smaller radius for quick turns. With that information, you can begin narrowing down your choices and make an informed decision rather than relying on marketing or a ski's paint job.

Ultimately, the problem of women's-specific gear is a problem of choices. How can the industry offer the apparel, gear, and equipment people need to enter the sport of their choice and progress? As brands move away from "shrink it and pink it," they're taking either a women's-specific or non-gendered approach to get to the same outcome—more choices and better-quality products that meet the diverse sizes, shapes, and needs of more people in sports, not just those who have been customarily read as "athletic," "outdoorsy," or cisgender men. We know the traditional male-centric product development process doesn't work for everyone. The design process for products shouldn't be driven solely by legacy issues like manufacturing processes, supply chain partners, or even marketing strategies. It should be driven by the real needs of real people.

And those needs—whether it's gear, training, or nutrition—aren't static or uniform. They morph along with people's lived experiences across the different phases of their lives, the most crucial being adolescence, pregnancy and the postpartum period, and the menopause transition. These can be turbulent periods, when hormones, bone health, and injury risk shift and your relationship to your body and sports can change drastically.

The next part of the book explores what exercise and sports look like during each of these life stages, how we can make sense of the unique challenges women face, and how we can equip women with the tools to make the best choices in pursuit of their active lives.

9.

The Phenom Years

•·•◦●◦•·•

I t's a rite of passage every spring in my neighborhood in Brooklyn, New York—opening day for the local youth baseball and softball rec league. Kids, decked out in their brand new uniforms, parade down Seventh Avenue and up to Prospect Park, where the first games of the season are played. It's mostly elementary school kids getting their first taste of playing on a team, and they're excited. At this age, sports are fun.

But as kids get older, the tenor around sports starts to change. Some of the joy begins to leak out and is replaced by competitiveness and a drive to win. Across the board, sports becomes a more serious endeavor, one that requires commitment and investment from kids and families. It's a phenomenon that Helen Tilghman has noticed over the years. Tilghman rowed in high school and college and has coached at the collegiate level. Now, as the head youth coach at Pocock Rowing Center in Seattle, Washington, she wanted to rethink their youth rowing programs, something the center has considered for a while. What does it mean to be a competitive youth athlete? How could they avoid getting completely sucked into the professionalization of youth sports, a trend they'd seen across the country? How could they—as coaches and as an organization—support long-term athlete development and keep sports fun?

It wasn't that they wanted to be less competitive—Pocock's youth programs have a reputation for being competitive on the national level—but they were aware of the potential costs associated with high performance athletic experiences for adolescents. They wondered if there was a way to avoid the overuse injuries and burnout they'd seen in so many kids throughout the sport. "I would rather have an athlete get tenth at Nationals their senior year and go on to win four national championships in college than have an athlete get first at Nationals [in high school] and not even make it through four years of collegiate rowing," Tilghman told me.

When you look at the structure of a typical youth rowing program in the United States, you can start to see some of the potential pitfalls. The earliest kids start rowing is middle school. With no real rec leagues and few school-based teams, most kids sign up with competitive club teams that run practices five to six days a week for nine to ten months out of the year. Tilghman explained to me that the traditional model in the sport is based on how long you've rowed, not how old you are. You start with what's called a novice year and then move up to the varsity team. "You could have some people that are doing their novice year as eighth graders, some that are juniors in high school, and then they all transition to the varsity team," she says. While athletes may be grouped based on their ability, they're all essentially doing the same amount of work. And that work typically involves high-mileage, high-volume training.

It takes a while to develop the physiology and aerobic capacity needed for rowing and for racing distances as long as 2,000 meters. Young athletes might think they're capable of making the jump to the next level of training and competition and want to make the transition, but their bodies may not be ready to do so, something Tilghman says most programs don't consider. She was concerned that asking an eighth grader or high school freshman to do the same level and intensity of workouts as a high school senior was unfair, and potentially dangerous. Plus, the training demands of the sport have increased over the years. "Collegiate programs are doing what the national team did ten years ago. The top junior programs are doing what collegiate programs were doing ten years ago," Tilghman says. "But the athletes aren't any older."

Overtraining and overuse injuries can crop up toward the end of high school or in the middle of a collegiate career, plaguing athletes and limiting their development and performance potential. High levels of burnout force some to retire at age eighteen, even though rowing is a sport where you physically peak in your late twenties. "It's a lot of people walking away," she says. Plus, rowing is expensive and requires a tremendous time commitment, which can make it exclusionary and unwelcoming for many kids and families.

Fundamentally, adolescence is a critical time in a youth's relationship with athletics and physical activity. Yet within youth sports, there's often a disconnect between the unique physiology of maturing kids—and what their bodies need during this time to be strong and healthy—and the increasing expectations of athletic training and performance in a system that's becoming more competitive. This is especially true for girls, whose needs are often misunderstood and overlooked. With little advice to guide girls through this stage of life, athletes, parents, and coaches are left to fumble in the dark in a way that may compromise a young person's long-term physical, mental, and emotional health, not to mention their love of sports.

Tilghman and other Pocock coaches recognize the shortcomings of the traditional model of competitive rowing. "As a young coach, you don't necessarily know what it is exactly you want to do but you certainly have a pretty good idea of what you don't want to do," she says. Quite simply, she doesn't want her athletes injured and she doesn't want them to hate their sport. Tilghman's concern transcends competitive rowing. It's a worry that I heard from coaches, parents, and young athletes themselves across multiple sports.

Could there be a better, safer experience within youth sports?

I f we want women to be happy, healthy, active people and athletes over their entire lives, we have to start by looking at youth sports—the system they're funneled into, the beliefs it fuels about bodies and what it means to be an athlete, and the culture that maintains it. The thinking goes something like this: If you want to be a good athlete, you need to practice, practice, practice. If athletic excellence depends on accumulating hours of training, kids should focus on one sport from a young age. Surely then they'd progress more rapidly, positioning them for a college scholarship or a spot playing at the elite level.

In gymnastics, for instance, young women are often put on pace to peak at age fifteen or sixteen, even though they may not physically peak until their late teens or early twenties. According to Georgia Cervin, author of the book *Degrees of Difficulty: How Women's Gymnastics Rose to Prominence and Fell from Grace*, the age of elite gymnasts plummeted in the post–World War II era in part because the sport began to favor a more acrobatic style that suited youthful, undeveloped bodies. When Nadia Comăneci dominated the gym at age fourteen, earning the first perfect 10 at the 1976 Olympics and winning three gold medals, including the all-around title, the model of "start young and specialize early" solidified. Tennis, too, is notorious for its teenage stars. In 1978, Nick Bollettieri opened his namesake tennis academy in Bradenton, Florida, where athletes lived and practiced full-time, pioneering (and professionalizing) a new paradigm for training top junior elite tennis players.

As generational talents like Tiger Woods, Michael Phelps, and Mikaela Shiffrin emerged, the belief in training younger and training more spread to other sports, including team sports, and it permeated the public consciousness after the publication of Malcolm Gladwell's book *Outliers* in 2008. Gladwell looked at the factors that make high-achievers succeed and described the "10,000-hour rule," the magic number of hours of deliberate practice required to achieve greatness. The idea stemmed from a 1993 study of violinists, which found that top-ranked musicians accumulated an average of 10,000 hours of practice by the time they were twenty years old. Gladwell then used this principle to explain the accomplishments of others like the Beatles and Bill Gates. While Gladwell also discussed the crucial role innate talent and opportunity play in the process, it's hard to ignore the 10,000-hour rule. It's seductive. Success depends on chipping away at a concrete number of hours—not luck, chance, or circumstance. (One of the study's authors, K. Anders Ericsson, has since said the study's findings have been misconstrued and oversimplified, that expertise isn't just about a specific number of practice hours.)

Today, the idea of sports specialization has morphed into an all-encompassing state of mind. Youth are likely to be funneled into a highly structured, professionalized system at a much earlier age. Neeru Jayanthi is the director of Emory's Tennis Medicine Program and one of the country's foremost experts on youth sports health. He says that roughly one-third of young athletes are highly specialized—those who quit all other sports to concentrate on a single discipline and who train more than eight months of the year. Whereas previously kids would choose one sport around fifteen years old, kids and parents now are making that choice between age nine and twelve.

As more kids and families travel down the specialized route, an entire ecosystem has emerged. "Adults have moved into what was a free play, unstructured space and started building these really elaborate structures," says Julie McCleery, an educator, researcher, and advocate who studies coaching and youth sports. These structures are built on the assumption that kids want the same intense workouts and high-stakes competitions adults enjoy, often nudging aside principles of positive youth development and age appropriateness. The current system can lose sight of sports as play and a vehicle for fun and friendship, for growing as a human.

It's led to the development of what McCleery calls the "youth sports industrial complex," an estimated $19 billion industry. Across the country, travel and club teams, talent-development programs, and sports academies are now staples, and you can find any number of tournaments and showcases taking place every weekend. These programs are designed to identify the best kids and move them up the ladder. Those who make it to the next tier are asked to commit more—time, training, and especially money, with families shelling out thousands of dollars because their kid could be "the next big thing."

It's created a dilemma for parents because there is intrinsic value in physical activity and sports. Sports were an important part of my adolescence, but it wasn't about winning or losing. (I wasn't a great athlete.) It was about being a part of something bigger than just me, something that brought me together with friends, something that taught me more about myself. I couldn't wait for my two kids to have a similar experience. While I wanted to give them a leg up, I wasn't sold on the idea of the single-sport track. But everyone was doing travel and club teams, and I worried they would be "behind" and unlikely to catch up to their peers if they didn't specialize. "[Parents] hear this message that more is better. If my kid starts younger and specializes earlier, she's going to get that scholarship," says Andrea Stracciolini, director of medical sports medicine at Boston Children's Hospital. Once you're committed to that path, Stracciolini says it's hard to disengage because "their child's athletic identity is in it. The parents' identities are in it."

The kicker is there's no definitive evidence that early specialization leads to future success. Youth coaches may not be incentivized to focus on an athlete's development and longevity, especially if they're evaluated on the team's or an athlete's current performance. Performing the same skills over and over means the same muscles, connective tissue, and joints—ones that aren't yet fully developed—are hammered continuously. The repetitive load can add up, especially when young athletes' neuromuscular networks are still under construction. Compared to unspecialized youth athletes, single-sport athletes are twice as likely to be diagnosed with a serious overuse injury that sidelines them for a month or more. Stracciolini says injury can be "devastating to their mental health" because athletic identity can be so essential at this life stage.

Experts say girls may be disproportionately affected since they are more likely to participate in individual technical sports like dance, figure skating, gymnastics, and tennis, which demand early specialization. Not only do individual-sport athletes specialize earlier, they tend to carry higher training loads. Even among girls who play team sports like soccer, volleyball, and basketball, those who only play a single sport experience knee pain more often compared to those who play multiple sports. The odds of sustaining a serious injury is more than two times higher in highly specialized soccer players than those who aren't specialized.

Lisa Joel, athletic director and head coach for the girls' varsity soccer team at Phillips Academy, an elite secondary school in Andover, Massachusetts, told me that injuries happened when she was a three-sport athlete at Amherst College, but it's nothing like what she sees today among high school athletes. In the current culture, athletes are expected to train year-round and six to seven days a week. Sometimes they work with private coaches or play with club teams in addition to their school teams. Joel has had to change her coaching model, paying more attention to the intensity of training her athletes are exposed to across a week and a season in order to create more opportunities for physical and mental recovery and avoid overtraining.

S ports and physical activity alone don't put young people at risk, nor is sports specialization itself completely to blame. Sports do require fitness and skill proficiency, and some disciplines, especially technical ones, demand a certain degree of commitment and focus at a young age to reach higher levels of competition. The problem is that the current model for youth sports lacks an understanding of the maturation process and what's appropriate for growing bodies and minds.

The preteen and teenage years are marked by rapid, sometimes turbulent, physical growth and hormonal fluctuations as kids' bodies reshape themselves. While the timeline for puberty is different for everyone, changes can occur seemingly overnight. Suddenly, kids inhabit a completely new form, one their brain is not used to controlling, which can alter their coordination, stamina, and motor skills. You see it during a growth spurt, when kids look like baby giraffes trying to get their legs under them, moving clumsily and without any notion of their body's position in space.

At the center of this transformation are bones. During adolescence, bone mineralization can't keep up with bone growth, leaving the skeleton temporarily more fragile. Think of it like the fundraising thermometer at your local YMCA. As bones elongate between age ten and fourteen, it's like drawing the outline of the thermometer. To reach maximum bone health, you need to color inside it. In this case, that means building bone mass and density to bolster the bone's interior structure and strength. Young people accumulate roughly 90 percent of what will be their adult bone mass by the end of adolescence; they reach their maximum bone density, filling in the entirety of their thermometer, between the ages of twenty and thirty. If they don't reach the top of the thermometer or maximum bone mass during this critical period, they can't add more bone to the bank. From that point on, they can only maintain what they've accumulated.

Boys generally emerge from puberty with a more "athletic" build. They're leaner and stronger, largely due to androgens like testosterone flooding the body, and they accumulate bone mass at a greater rate compared to girls. There's a natural evolution of strength, power, and coordination, says Greg Myer, director of the Emory Sports Performance and Research Center, a "neuromuscular spurt" that accompanies the growth spurt, and boys' athletic progression builds step-by-step. In his research, Myer found that not only did boys jump higher as they matured, they also knew how to control their bodies upon landing.

Girls go through a similar period of explosive growth during puberty, but they encounter more hazards that can get in the way of building strong bones. While genetics, body size, race, ethnicity, and family history all influence bone health, hormones like estrogen, which are released during menstruation, are essential to building and protecting bone mass. That means the age at which a young person begins to menstruate and the regularity of the cycle can affect long-term bone health. If the first menstrual cycle doesn't begin by age fifteen, it's like being stuck at zero while everyone else gets a head start coloring in their thermometers. If a young person experiences absent or infrequent periods, it's like coloring with a dried-out marker. Instead of bold, dark strokes, the thermometer looks faded and patchy. Those with menstrual irregularities are two to four times more likely to suffer a stress fracture and are at a greater risk for osteoporosis at an earlier age.

What's more, young athletes are more likely to start their periods later and report menstrual irregularities compared to nonathletes. Not only do they have lower bone mass density, they also have weaker bone tissue. The trabeculae—the microscopic struts that act like scaffolding in bone tissue and give it its spongy appearance—are fewer and farther between, making the bone less structurally sound.

Muscle growth doesn't keep pace with skeletal development either. Myer says it's the equivalent of giving girls a bigger car but not giving them a bigger engine or the "neuromuscular horsepower" to play at this higher level. For many, athletic progression stalls instead of moving in lockstep according to the straightforward trajectory seen in boys and men. While hormones may play a role in these changes, Myer told me it's also likely that girls don't get the same strength and conditioning opportunities that boys do. They're not taught that strength training will build the resiliency they need to progress in their physical activity and sports.

It may explain why girls' biomechanics during dynamic movements start to diverge from boys' around puberty, such as their tendency to land with a more knock-kneed position, a red flag for potential injury. They don't develop the same neuromuscular control as boys to stabilize their longer limbs, wider pelvis, and greater body mass. Ultimately, the combination of physical, neuromuscular, and biomechanical changes can send more stress through the major joints. It creates what Andrea Stracciolini calls a "perfect storm for significant injuries," ranging from acute injuries like ACL tears to chronic conditions in the hips, pelvis, and spine—injuries that are mostly absent in prepubescent girls.

The implications of these tremendous changes are largely ignored and misunderstood within youth sports. Adolescence is a liminal phase when the body is more pliable and less resistant to physical stress because it hasn't matured into its adult form. It's also the time when youth are asked to increase the hours dedicated to sports and the intensity of their workouts. Young athletes want to advance their skills too. They want to up the ante of their tumbling routines in gymnastics, run more mileage, and play more tournaments. They want to pitch more. But by and large, coaches aren't monitoring a girl athlete's training volume in relation to her growth and development. "Nobody counts tumbling passes in the growing gymnast, but they count pitches in the young baseball player," an area where there's "droves and droves of research," Stracciolini says. "That's a disservice."

Most athletic programs and sports organizations don't always know how to guide girls through this transition or how to balance athletic development and performance in the near term with long-term well-being and longevity in sports. Historically, awareness of menstrual health and conditions like the female athlete triad and RED-S has been low. Some studies showed that less than half of Division I coaches and only 14 percent of high school coaches could name all three components of the triad, and less than half of physicians, athletic trainers, and physical therapists could identify those same components. More recent studies at the collegiate level suggest that knowledge has improved, but experts say that more education is still needed. If coaches and healthcare professionals aren't attuned to these conditions, how can they pass information to girls and parents? "That's a travesty. We've known this for years and people still have no idea," Stracciolini says.

Even breast health and the necessity of sports bras is overlooked. Nearly three-quarters of young people with breasts reported at least one breast-specific concern related to sports and exercise—they're too bouncy, too big, too small—but only 10 percent always wear a sports bra. It's no wonder girls who mature early are less likely to take part in physical activity, and 44 percent of teen girls don't participate in sports at all.

Instead, there's a perception that there's something wrong with a growing and developing young woman's body that makes it incompatible with and counterproductive to sports, and the body becomes a problem to be fixed. Girls fight their physiology by training countless hours and undereating to stave off the inevitable changes that accompany puberty. Many believe that losing their period is normal, a sign their body is fit rather than a signal there's something wrong. They're prescribed birth control pills to induce a menstrual cycle, which doesn't address the underlying nutritional and hormonal issues that cause cycle irregularities. Or they (and their coaches) buy into outdated notions like "thinner is faster" because for an alluringly brief period of time, it seems to work. As weight plummets, so do race times, until the young athlete herself falls off the map. There are also social and cultural pressures to contend with—peers teasing girls who play sports, parents believing boys are better athletes than girls, and girls themselves feeling self-conscious and awkward about their changing bodies.

The failure to truly consider how adolescents develop and the misplaced focus on performance expectations robs girls of health, their potential for success, and the joy of physical activity and its wide range of benefits. Overall, girls drop out of sports at rates unmatched by boys, and those rates accelerate with each successive grade in high school. By the age of seventeen, 51 percent of girls have left the athletic arena. And for girls of color or lower socioeconomic status, these disparities are even more pronounced.

When she was a collegiate runner at Wesleyan University, Yuki Hebner's coach gave her the "fat talk," telling her she had the potential to be an All-American only if she lost weight. Without guidance from trained nutrition professionals, Hebner was left to shed pounds on her own. She resorted to a combination of high-volume training and underfueling—pervasive practices on the team—and purging. Ultimately, she sustained a femoral stress fracture. She says, "When my body began changing at nineteen, I was led to loathe and harm my adult body rather than learning how to adjust and harness strength from it."

Olympian and professional distance runner Kara Goucher felt similarly when her body started to morph during her junior year of high school. She grew taller, thicker, and softer and got her period for the first time. Running had always been easy, but her body stopped responding when she pushed. Goucher felt like she was failing. "I'd been told for all these years how talented I was, and I just felt like I somehow got lazy and got heavier and taller," she told me. Her self-esteem took a hit. She told me that the changes she experienced during puberty "completely changed the trajectory of where I went to college" because she no longer ran fast. Instead of setting her sights on becoming the best runner in the country, she focused on finding a good collegiate team, one that "someday I could make the varsity team and score for," she says.

The changes girls experience during adolescence—a time when teens become more body conscious and start dieting—can ignite disordered eating and eating disorders precisely when they need more energy and nutrients. Comments from coaches or others about bodies and eating habits have been found to trigger body dysmorphia and disordered eating issues that women continue to struggle with years later.

It takes time to grow into your body and adjust to the tumultuous changes. Though Kara Goucher felt that her body had betrayed her, it hadn't. It was doing what it needed to do to get her to the next level in running. Halfway through college, she hit her stride again. While home in Minnesota for Christmas break her junior year, Goucher ran the paths along Lake Superior. When she got home, she cried. Her body no longer fought her and she felt connected to it in a way that had eluded her for years. She realized, "I'm still in there." After returning to campus at the University of Colorado Boulder, she qualified for the NCAA indoor track meet, a goal she thought wasn't possible for her collegiate career. "I was no longer looking back, wondering, 'Could I find that person again?' It was all forward looking from then on," she says.

When I talked to parents about youth sports, many told me that while they hoped their kids would be good athletes, they ultimately just wanted them to be happy. That's why Carol Vaasili Brice's thirteen-year-old daughter Leilani liked playing with her travel soft-ball team. Her teammates were her friends and they had fun together. As the team grew from 8U to 12U, players settled into specific positions, and Leilani became the team's third baseman. Vaasili Brice started to see her daughter's relationship to the sport change. Leilani wanted to try different positions like first base, but outside of a game or two, she wasn't given the opportunity. Leilani used to ask her mom to play catch or practice pitching with her, but Vaasili Brice told me, "I don't think I've heard that in at least a year and a half, two years."

When Leilani started playing club volleyball, Vaasili Brice saw an excitement in her daughter that was missing from softball. It made her question whether she should have urged her daughter to take a break from softball sooner. "We try to convince ourselves that their experi-ence is going to be our experience," she told me. But it's different today. While sports was Vaasili Brice's only outlet, her daughter has multiple enrichment activities—art, music, Science Olympiad—and sports doesn't have to be her everything.

Lisa Joel at Phillips Academy sees less joy in kids playing sports. When you're a club athlete who plays multiple games a weekend or your sole goal is to secure an athletic scholarship, sports becomes transac-tional. Kids who as freshmen told her they want to be collegiate athletes have no interest by senior year. They miss playing sports for the love of it, when the stakes aren't so high. "They're not happy," she told me, which contributes to burnout.

The tiered system of youth sports also pushes out kids who can't afford to level up to travel teams or sports academies, who come to sports late, or who just want to participate for the enjoyment of it, like distance runner Alexi Pappas. Growing up in Northern California, Pap-pas was one of the top teenage runners in the state, but as soon as running became intense and serious, it was no longer fun. Her high school athletic department believed Pappas needed to remove all other distractions—soccer, theater, student government, and her bustling social life—and her coach gave her an ultimatum junior year: quit soccer or get kicked off the cross-country and track and field teams. Pappas wasn't ready to dedicate herself to the singular pursuit of running just because she was good at it. She wrote in her memoir *Bravey*, "I was gradually growing into the sport just as I was gradually growing into myself." Instead of specializing, she stopped running.

It's indicative of the "varsity or bust" mentality. Joel says if a tenth grader doesn't make the varsity team, they don't see the value of playing on the JV team. It trickles down to younger kids too, where the pay-to-play system serves as a gatekeeper to high school sports and there isn't the same opportunity to walk on to a team or play multiple sports like in the past. Jenny Martin noticed things change when her daughter Lila was in sixth grade. Rec league teams started to fall apart as girls either dropped out of sports or stepped up to a travel or club team. At the time, Lila's softball team blended the competitive and rec teams and the difference in skill level was palpable. Inevitably, the girls who played rec-only didn't get to pitch, catch, or play infield positions and had limited playing time. "Softball was fun up until it stopped being fun," Martin told me, and Martin wanted Lila to enjoy sports, not burn out on them like she had.

Contrary to the dogma spawned by the book *Outliers*, a recent study of more than six thousand athletes found that those who had a head start as kids were more likely to perform well at the youth level, but didn't see the same success as they got older. Instead, those who achieved world-class status at the senior level participated in multiple sports as a kid, started their main sport at a later age, and progressed slowly and deliberately. In other words, they didn't specialize early. They weren't teenage phenoms. Or as journalist and author David Epstein writes in his newsletter, *Range Widely*, "If the goal is to win the 10-year-olds' championship, then early specialization seems to make sense. If the goal is to develop the best 20- or 30-year-olds, then not so much."

When we zoom out, we see that the biggest gift we can give young athletes is space and patience to develop according to their own timeline. We see that when athletic progression begins to dip and kink, it's temporary rather than a harbinger of future failure. The line will naturally unknot itself—if we help kids ride out the turbulence of adolescence and young adulthood, encouraging and nurturing them so they don't give up on sports and physical activity altogether, and if we remind them to have fun.

While her coaches thought she'd lose her potential to "the vortex of puberty," Alexi Pappas writes that her "forced retirement from high school running became a major advantage in my later growth as an NCAA and then professional athlete. I inadvertently stopped training just long enough for my body to go through puberty." There wasn't pressure to underfuel in the name of maintaining the "twisted Peter Pan prepubescent body" many believed was the key to success as a runner. She didn't overtax her musculoskeletal system when it was at its most vulnerable. What appeared on the outside as a fallow, idle period was actually an opportunity for her body to regroup, to build the foundation for a more durable and resilient form capable of handling the increased training required as a collegiate, professional, and Olympic runner.

Prioritizing physical and mental health over the high-stakes performance expectations of youth sports programs is paramount to long-term health and athletic success. We can't assume that what works for adults will work for kids. While Pappas's break from sports was fortuitous, we can be more intentional in respecting the maturation process and helping youth adapt to their changing bodies rather than forcing them to conform to a one-size-fits-all developmental pipeline.

A better approach to youth sports would focus on building a diverse range of fundamental motor skills along with an enjoyment of movement through free play and sampling multiple activities. A varied athletic background helps kids develop neuromuscular flexibility and may better equip them to learn new skills. After Emma Raducanu won the 2021 U.S. Open, her first Grand Slam singles title at age eighteen, her coach Matt James told *The Guardian* that Raducanu's exposure to multiple sports—ballet, horse riding, swimming, basketball, skiing, and golf—before focusing exclusively on tennis was an advantage. "When she's learning a new skill, or trying something a little bit different, she has the ability and coordination to pick things up very quickly, even if it's a quite a big technical change," he said. Even Serena Williams tried a variety of sports like gymnastics, track and field, tae kwon do, and ballet.

A late start turned out to be a secret weapon for world champion mountain biker Kate Courtney. She started racing bikes at age fifteen, which she says was about "ten years too late." She had to play catch-up with her European counterparts, and she told me she got her ass kicked. Since she wasn't expected to do well, she focused on dialing in the fundamentals. Over her four years of competing on the Under-23 World Cup circuit, she steadily improved her year-end standing from eighth in her first year to first in her last year. When she reached the elite level, her team took a similar measured approach. "We prioritized performance and putting on power not just for this year but for the longevity of my career," she says.

Courtney also chose to attend Stanford University, deliberately slowing her progression. "People underestimate the pressure and challenge of being a full-time athlete. That wasn't something I was really ready for until I graduated college," she says. Mountain biking wasn't her entire life or identity, and that protected her from burnout. By the time she graduated and went pro full-time, she was physically, mentally, and emotionally mature. In her first two years competing at the elite level, Courtney captured the 2018 UCI Mountain Bike World Championship title, becoming the first American to don the coveted rainbow-striped jersey in seventeen years, and the 2019 Overall UCI Mountain Bike World Cup title. In 2021, she competed in the Tokyo Olympics.

Many young people will choose to specialize, even though current research recommends not zeroing in on a single sport until middle or late adolescence. In order to navigate these waters, they need guidance to keep them safe. Generally, experts say total weekly training hours shouldn't exceed a child's age. If a kid is twelve years old, the total amount of time she spends in training, practice, and games shouldn't be more than twelve hours in any given week.

Coaches and sports programs can help young athletes sidestep injury and burnout by carefully monitoring their growth and maturation alongside their enjoyment of sports and adjusting their training load as needed. Take girl gymnasts, who often go through a period during puberty where they have trouble performing skills that used to be routine. They can become temporarily confused in space or lose their timing as their limbs lengthen and center of mass shifts. If a gymnast is expected to barrel through workouts without modifications, it can breed frustration, low confidence, and injury. But if coaches adjust the amount and type of work an athlete takes on during vulnerable periods, they give them a chance to recalibrate and reconnect with their bodies.

Coaches can create individualized strength training and conditioning programs designed to shore up weaknesses at different stages of maturity too. Encouraging athletes to focus on developing gross motor skills and strength before they enter puberty can help prevent the changes in biomechanics and neuromuscular control from being as disruptive. According to Greg Myer, this is particularly important for growing girls, who are more susceptible to injury during this time.

Kids, parents, and coaches should also prioritize recovery—it's when the magic happens. Kids need time off from organized physical activity every week, a bare minimum of one day. Experts recommend taking a month off after a season ends and a total of three months off during the year. Stepping back from competitive sports gives the body and mind a chance to relax and recoup, creating a buffer against burnout and overuse injuries.

Recovery is more than just time spent not exercising. Stress is stress, whether it's physical stress from workouts or mental stress from homework or emotional stress from a bad breakup. It accumulates in the body. While there are endless gadgets and supplements designed to facilitate recovery, the most important tool for youth (and all of us) is quality sleep, especially since more than 70 percent of high school students don't get the recommended eight to ten hours. Muscles, bones, connective tissue, and the brain repair themselves during sleep, and the body consolidates training adaptations to improve strength and performance.

In the fall of 2020, Pocock Rowing Center began phasing in youth programming based on age rather than experience. They now have a middle school program and a separate high school program with three age groups—under-17, under-18, and under-19. It allows them to create tiered training based on the rowers' physical and social-emotional development. They've also changed their schedule so they're no longer forcing kids to commit full-time from the start. Middle school rowers practice three days a week and the commitment scales up as they get older. To encourage kids to play multiple sports and to take time off, they've adjusted their season so that it doesn't directly conflict with the school sports schedule.

Head coach Helen Tilghman said the change was a hard sell for some parents who worried their kids wouldn't be as competitive, but she believes that instead of "going for broke in any given moment to be as competitive as possible," her athletes will be better and healthier over the long term. She already sees glimmers in the kids who showed up consistently even when competition paused during the COVID-19 pandemic. Their foundational skills are strong because "there was no ticking clock on getting them to race speed," she told me. "We're already seeing those athletes doing incredibly well on our team. In the next year or two, I think we're going to see those dividends really increase exponentially." She says more rowing coaches want to see a shift toward an athlete development model that prioritizes safety and longevity.

Pocock is following in the footsteps of the Women's Tennis Association (WTA), one of the first organizations to match the professional and physical demands of sports with an athlete's age. Beginning in the 1970s, teenage phenoms as young as thirteen years old shook up the WTA tour with their precocious talent and set a slew of "youngest ever" records. But Emory's Neeru Jayanthi says these players were asked to do things "without consideration for the fact that they were still children." Physically, mentally, and emotionally, they weren't prepared for the demands thrust upon them; they were pushed too hard too soon. Tracy Austin, former number one in the world and youngest U.S. Open champion at age sixteen, retired at age twenty-one. Andrea Jaeger, former number two in the world, lost her passion for the sport and retired at age nineteen due to injuries. The anguish and pressure of early success led three-time Grand Slam champion and Olympic gold medalist Jennifer Capriati to take a fourteen-month break from the sport at seventeen, after only four years on tour. She told *The New York Times* in 1994 that she was depressed and lonely, that she thought she was a "loser" if she didn't win.

In 1995, the WTA instituted an age eligibility rule to ease young athletes into the world of elite tennis. Beginning at age fourteen, athletes can play in up to eight professional tournaments; the allotment gradually increases each year until age eighteen, when restrictions are lifted. Athletes must take part in WTA's Player Development programs, which helps them assimilate to life on the tour and navigate the stressors of professional sports—injury, media, travel, schedule planning, and long-term career development. A ten-year evaluation of the age eligibility rule and professional development programs found fewer athletes retired prematurely (before age twenty-two), declining from 7 percent before the rules were instituted to less than 1 percent after. Now, players make their professional debut later and the median career length has increased by 43 percent.

Others are building a more nuanced approach based on practices like age eligibility rules. Since the timing and tempo of development can vary greatly, it's not uncommon to have a group of athletes who are the same age but on opposite ends of the maturation spectrum. During a period of rapid change like adolescence, these differences could span two to three years, and young athletes may need different training environments.

It's not a new concept. In the early 1900s, people questioned whether chronological age was the best way to determine if a child was ready for factory work. In 1904, physician C. Ward Crampton proposed the idea of "physiological age," based on the development and characteristics of a boy's pubic hair. Soon, the notion of grouping kids based on maturation status moved from factories and into school and sports. In 1916, the Baltimore Public Athletic League noted, "It would be justifiable to arrange physical training schedules in schools on the basis of physiological age, giving boys or girls of the same physiological age similar types of exercise." Boys were again evaluated based on pubic hair development, while girls were assessed according to menstrual cycle status and breast development. By midcentury, it was clear that physical maturity was advantageous in sports, particularly for early adolescent boys. Fifty-five boys who competed in the 1957 Little League World Series were surveyed and more than 70 percent were considered mature for their age.

While youth sports are often organized based on birth year, grouping by maturity status hasn't been implemented as widely. Recently, there's renewed interest in the idea and how it could support athlete development by providing a rich sporting environment for both early and late bloomers. In particular, the concept of bio-banding has gained more attention—where kids are categorized based on their chronological age and maturity status, which is determined by how close they are to their predicted adult height. It's typically implemented between the ages of eleven and fifteen. Sports programs and leagues evaluate youth at the beginning of a season and monitor them throughout. Those determined to be less than 85 percent of their predicted adult height are considered prepubertal, while those between 85 and 89.9 percent are categorized as early pubertal. Those between 90 to 94.9 percent are classified as midpubertal, kids who are about to enter or are in the midst of their growth spurt. Youth who are 95 percent or more of their predicted adult height are considered postpubertal and close to skeletally mature.

With these four bands in place, coaches can move kids to older or younger training and competition groups based on an athlete's physiological parameters, as well as their psychological readiness and technical skills. The hope is that in doing so, coaches and leagues can reduce some of the mismatch in size, strength, and power you see when early maturing and late maturing athletes play together. For youth experiencing rapid growth, coaches can help them safely move through this period and reduce injury by decreasing training volume and intensity while shifting the focus to strength training and neuromuscular development.

In recent years, the English Premier League piloted this approach with their boy soccer players. U.S. Soccer has followed suit, expanding the program to include both boy and girl players. What they found was that bio-banding levels the playing field. Those who developed early couldn't depend on physical speed and strength to dominate the game. When they played opponents matched for maturation, they had to develop a well-rounded game and rely on a more technical and tactical approach. Late bloomers were no longer penalized or overlooked for their smaller size and had the opportunity to play to their full potential. When playing with younger players, they could step up as leaders for the squad.

"If you give [athletes] time to develop, if you give them the opportunity to have an individual environment to succeed, then every player, whether they're playing on a national team or in grassroots [amateur sports at the local level], will be in the best environment to develop," concludes Tom Hicks, former senior manager of U.S. Soccer's High Performance Operations.

While policies like age minimums and bio-banding are a step forward in recognizing the developmental needs of young athletes, the realignment of youth sports toward the whole athlete and their overall health can't happen without a cultural shift: less judgment based on the number of titles, records, and championships an athlete does (or doesn't) garner, more frank conversations about the physical, mental, and emotional needs of girls and young women. In short, less treatment of athletes as disposable and replaceable. More *fun*.

Today's young women are leading the charge. They aren't waiting for schools, colleges, or sports governing bodies to do the right thing. They're pushing back against the systems that prioritize winning over the unique needs of young people, demanding more—and better— from the sports they love, from those in charge, and from the system within which sports exist.

A new wave of athlete empowerment is rippling across the globe, unleashed by the Larry Nassar gymnastics abuse case, by Mary Cain's experience as a Nike runner, and by the wider #MeToo and Black Lives Matter movements that brought social justice issues to the forefront in recent years. In 2021, seventeen former gymnasts filed suit against British Gymnastics, alleging that "British Gymnastics implemented a model of suspended pre-pubescence leading to generations of girls with eating disorders, body image issues, and deliberately stunted physical development." Current and former soccer players in the United Kingdom spoke up about the obsession with weight and body image that pervades many clubs in the Women's Super League, fueling a whole host of physical and mental health problems.

Collegiate athletes have mounted the most powerful displays of resistance, standing up to say that the genuine lack of concern for athlete welfare is not acceptable. In March 2020, thirty-six cross-country and track alumnae from Wesleyan University, led by Yuki Hebner, published an open letter and testimonials alleging that the toxic culture on Wesleyan's running teams condoned disordered eating and body shaming under the false pretense that a leaner, skinnier build led to better performance. They said it was a mentality that "recklessly compromised our health in the process," resulting in injuries and high attrition rates.

What was unique about the Wesleyan effort was that the group didn't want to oust the head coach. They wanted the university to address the systemic issues that aided and abetted the practices that crippled young athletes. They called for qualified medical professionals to provide guidance on nutrition, bone health, and menstrual health rather than leaving it to the coaching staff. They asked the university to designate an athletic trainer with expertise in endurance sports for the racing teams (cross-country, track, crew, and swimming), providing them with the same standard of care as sports like football and basketball. They requested that the university commit to hiring more women running coaches, coaches who can act as role models for student-athletes and who may relate to the lived experiences of those on the team. They also demanded the university revamp its coach evaluations to include issues such as attrition, injury, and athlete health and proactively identify detrimental coaching patterns. (Since the letter surfaced, the head running coach retired, and Wesleyan committed to implementing additional measures to support athlete health.)

In the months following the Wesleyan letter, athletes on the University of Arizona's women's cross-country and track teams went public with their own stories of harmful and negligent practices, which included overtraining, bullying, and body surveillance. Between February and March 2021, more dominoes fell. Current and former student-athletes at the University of Alabama at Birmingham, Bradley University, Pepperdine, and Loyola Marymount University (LMU) came forward on social media with similar accounts of pervasive toxic cultures within their school's running teams, including some young men athletes. At the University of California, Berkeley, current and former athletes broke their silence about the history of fat-shaming, intimidation, and mistreatment within Cal's storied women's soccer program.

In the past, young women suffered alone, often feeling personally responsible for their struggles. Yuki Hebner wrote, "My teammates and I were incapable of supporting each other through changes we did not understand and struggles we could not acknowledge." Similarly, LMU runner Rosie Cruz was led to believe that injuries, especially freshman year, were par for the course. She broke her foot during her first cross-country season. She broke it again during track season later that year. After each injury, she rushed back to high-mileage training, which she felt she had to do because it was her coach's training philosophy and she couldn't push back. Cruz visited several doctors and physical therapists, asking the same questions: What was wrong with her? Why couldn't her body withstand the mileage? They told her there was nothing wrong with her. The mileage was the problem.

These young women pointed out that it's not the problem of one athlete, one coach, or one school. It's a systemic failure of sports that leaves young women as collateral damage because coaching practices, training and injury prevention protocols, and social norms largely ignore the physiology and needs of young people who are still maturing. "It's incredible what happens when you are intuitive with your running and when you believe in yourself and listen to your body," says Cruz, who has been running happy, fast, and injury-free since stepping away from the LMU athletics program. After Cruz shared her story on social media, she was inundated with more than a thousand messages from athletes across the country who all shared similar stories of mistreatment. It showed that young athletes need more support, education, and resources to navigate the treacherous landscape of young adulthood and the relentless competitive environment.

Stef Strack wanted to give young girls the tools needed to change sports—community, a support system, and a voice—tools that she didn't have as a young athlete growing up as a ski racer and soccer player in Alaska. In 2019, Strack launched Voice in Sport (VIS), a company centered on girls ages thirteen to twenty-three, a demographic that drops out of sports at high rates. Through VIS, girls can connect with one another and collegiate and pro athlete mentors and schedule group or one-on-one sessions with more than eighty experts in women's health, nutrition, sports psychology, and sports science—resources young girls often don't know about and don't have access to outside of specialized sports medicine programs. These services can be a game changer. When girls are armed with the knowledge of what's happening in their bodies, rather than myths and misconceptions, they can identify potential problems and advocate for themselves.

Ashleigh Barty paused her meteoric rise to becoming a top-ranked tennis player in 2014 at age eighteen, after only three years on the WTA circuit. To outsiders, Barty appeared to be on the brink of greatness, but like Jennifer Capriati before her, Barty was unhappy and tennis had turned into a slog. She told reporters, "I needed time to step away, to live a normal life, because this tennis life certainly isn't normal. I think I needed time to grow as a person, to mature."

While Barty's two-year hiatus cost her tournament wins in the short term, it may have led to longer-term success. Three years after returning to the WTA tour, she won the 2019 French Open, her first Grand Slam title, and earned the number one ranking. Barty took a second break during the COVID-19 pandemic and came back to win the 2021 Wimbledon and 2022 Australian Open titles. In March 2022, the number one player in the world announced her retirement from tennis at the age of twenty-five on her own terms. "I am so happy and I'm so ready and I just know at the moment in my heart for me as a person, this is right," she said in her announcement on Instagram. She has so many more dreams she wants to chase, off the tennis court.

Track and field athlete Colleen Quigley told me we need to move beyond the point "where we place all of our value as humans on what we do athletically." Sports shouldn't strip young people of joy or leave some of the most talented athletes, like Naomi Osaka and Simone Biles, questioning their self-worth and deciding whether to sacrifice their health in the name of trophies and medals. You see the difference Quigley's making in sports at high school cross-country camps where girls (and boys) line up to ask her how to handle stress, disappointment, nerves, bad races, and comparison to others. You see it in the empathetic, sometimes cathartic, comments on social media when Quigley shares her frustrating battle with injury and the decision to pull out from the 2021 U.S. Olympic Team Trials for track and field.

Young athletes are making the bet that if they take care of their whole selves as humans, they'll be happier, healthier, and perform better. They're reimagining a system of sports that prioritizes health and longevity over short-term performance and excessive competition, a system that gets back to the core values and joy of physical activity and all it has to offer. In the process, they're modeling a new path forward for all athletes, girls and boys, young and old. "I'm making choices now that I feel are going to be really beneficial for the long term," Quigley says. "I can double down on myself and feel like I'm going to make choices that are best for me."

And she says it feels amazing.

10.

Family Matters

· · ⋅ ● ⋅ ● · ·

Aliphine Tuliamuk needed a minute. In February 2020, the distance runner for Northern Arizona Elite (NAZ Elite) convincingly won the U.S. Olympic Team Trials for the marathon on a relentless, hilly course in Atlanta, Georgia, securing a bid to compete on the biggest stage in global sports. A month later, the 2020 Tokyo Olympics were postponed due to the COVID-19 pandemic and Tuliamuk needed time to process the news. It wasn't just the disappointment and uncertainty of waiting an extra year before she could realize her dream and contend for a medal. It wasn't the inconvenience of rejiggering her training cycle, one that's precisely planned and timed to ensure she's healthy and in top form on race day.

For Tuliamuk, the fallout from the postponement reverberated well beyond the sporting arena. She and her partner, Tim Gannon (now her husband), wanted to start a family after the Olympics—now those plans were up in the air too. As she told *Women's Running*, she had to make a choice: have a child and go to the Olympics in less than peak form, or put off starting a family for another year and a half, which she said "feels like an eternity." Ultimately, she and Gannon gave themselves a two-month window to conceive. If she didn't get pregnant within that time frame, they'd try again after the Olympics. If she did, Tuliamuk would have roughly six months to recover from childbirth and get race ready. In the end, she got an extra month; her daughter, Zoe, was born in January 2021, seven months before the women's Olympic marathon.

Tuliamuk isn't the only athlete who has grappled with the complicated calculus of if and when to start a family. It's an equation that involves timing a pregnancy so it aligns with the Olympic, world championship, World Cup, or other major event cycle as well as weighing the longer-term ramifications of childbearing, childbirth, and the postpartum period on an athlete's body and career. Dawn Harper-Nelson, Olympic gold and silver medalist in the 100-meter hurdles, has said, "I've always known I wanted to be three things: a wife, a mom, and a gold medalist." But realistically, she knew it was impossible to do all three well. She was nervous that the physical and financial repercussions of motherhood would hold her back from being the best in her sport. Before having her first child, five-time Olympic beach volleyball player Kerri Walsh Jennings was told that pregnancy would affect her endorsements along with her body, that people would forget about her if she stepped away from the court to start a family.

For elite and professional athletes, their body is their job. Since prime childbearing years largely coincide with the years women athletes are at their peak, motherhood was never really considered part of the larger narrative for them. Rather, it signaled the denouement of an athletic career. Walsh Jennings described it as an "either/or proposition," where women are forced to choose between sports and family, with very real repercussions. If they put off having children, they may miss an opportunity to start a family. If they choose to retire and get pregnant, they may prematurely cut short their athletic careers. Typically, athletes have opted to wait until they retire to start a family.

If athletes do have babies midcareer, their experience during pregnancy and the postpartum period can fundamentally alter their relationship to sports and exercise. There are no solid protocols to help them safely navigate the return to physical activity after giving birth—it's simply assumed that the body will heal and bounce back on its own, since pregnancy and childbirth are viewed as "natural" processes. Elite athletes are left alone to manage the myriad physiological changes while also contending with contractual obligations and the implicit pressure to return to form as quickly as possible. If they rush back to sports without adequate support, they may experience injury or longer-term health problems. For some, the weight of these expectations can be what former pro runner Phoebe Wright called the "kiss of death for a female athlete."

And it's not just elite athletes who suffer.

Part of the reason family planning decisions for all active women are so fraught is because women are making choices in the absence of good information. There isn't enough data on exercise in active pregnant and postpartum people to make truly informed decisions. What activities are safe to perform and at what intensity? How does physical activity impact postpartum recovery, breastfeeding, and long-term health? People are left to make consequential decisions based on anecdotes, assumptions, and best guesses.

When it comes to women and their bodies, there are countless opinions about what they should or should not do. And when it comes to pregnant women and their bodies, those opinions are relentless. Drink more water. Eat this. Don't eat that. Take this vitamin. The most strident—and contentious—advice revolves around exercise, and women are bombarded with warnings about what is and isn't safe. These views, however, often are more reflective of, and fluctuate with, sociocultural norms rather than scientific evidence.

Historically, scientists were hesitant to include pregnant people in research studies because expectant mothers complicated the research process. Scientists would have to account for the vast anatomical and physiological changes that accompanied pregnancy, not to mention a number of research design and ethical issues. They were particularly concerned about maternal and fetal health and safety, especially after two high-profile incidents in the 1960s and 1970s.

The first case involved the sedative thalidomide, an over-the-counter medication available in more than forty countries and advertised as safe for everyone. Aside from its on-label use, some doctors found that it helped with morning sickness. But among those who took the drug during pregnancy, there was a rise in birth deformities, the most common being short, flipper-like limbs. On the heels of the thalidomide disaster, doctors discovered cases of a rare vaginal cancer in the daughters of women who, while pregnant, took diethylstilbestrol, a synthetic hormone that was prescribed to women in the 1940s and 1950s to prevent miscarriages.

No one would argue that there shouldn't be concerns about the participation of expectant people in clinical studies, but these incidents also created a strong aversion to the idea of pregnant people, and to a certain extent any woman of childbearing age, taking part in any scientific study. The federal government instituted regulations that effectively barred this population from research and drug trials. They were considered too "vulnerable," and further learning about the pregnant body came to a halt.

When prenatal exercise programs and guidelines emerged in the United States in the twentieth century, they weren't based on a solid foundation of research on exercise and pregnant people. The American College of Obstetricians and Gynecologists (ACOG)—considered the expert body when it comes to maternal and fetal health—didn't release their first prenatal exercise guidelines until 1985. It was the height of the fitness boom and aerobics craze and women wanted to know if it was safe to continue with their Jazzercise class or Jane Fonda workout videos with a baby on board. Previous recommendations from the U.S. Children's Bureau, issued in 1949, recommended that pregnant people continue only with housework, gardening, daily walks, and the occasional swim. Anything more strenuous was off-limits. ACOG, on the whole, endorsed aerobic activities for expectant people but warned against taking part in more than fifteen minutes of high-impact, strenuous exercise at a time. They also advised that women keep their heart rate below 140 beats per minute and core temperature under 100.4 degrees Fahrenheit.

While the ACOG guidelines were presented as black-and-white advice, they were largely based on the opinion of a committee of doctors. For example, the committee settled on the ceiling of 140 beats per minute because it seemed like a reasonable measure for moderate-intensity activity, not because maternal and fetal studies established it as a definitive threshold for safe physical activity. In erring on the side of caution, these guidelines seemed to suggest that there could be something inherently dangerous about exercising while pregnant. During my first pregnancy, despite knowing that regular exercise was good for me and could ease labor and delivery, I still worried I'd somehow harm my baby. I was cautious when running and lifting weights, especially once my belly popped. I mostly practiced yoga because my prenatal yoga class felt safer than the cardio or weight room at my local YMCA, where I got judgmental looks whenever I picked up dumbbells or ran on the treadmill.

Since the ACOG guidelines were released in the 1980s, more studies emerged demonstrating that exercise is beneficial and safe for both pregnant people and their babies. According to Michelle Mottola, maternal health expert and professor at Western University in Canada,

"There has been an explosion of evidence" over the last decade evaluating the impact of physical activity and exercise during pregnancy and the postpartum period. This research has helped peel back the medical community's cautious, protectionist recommendations. Australia and New Zealand, the United Kingdom, the United States, and Canada all published updated recommendations between 2016 and 2019.

"Exercise during pregnancy, unless you have a medical contraindication or a medical reason why you shouldn't exercise, is safe and beneficial. Full stop," Margie Davenport, associate professor at the University of Alberta, who led the development of the new Canadian guidelines, told me. As she and her colleagues wrote, the guidelines represent a "foundational shift in our view of prenatal physical activity from a recommended behaviour to improve quality of life, to a specific prescription for physical activity to reduce pregnancy complications and optimise health across the lifespan of two generations."

In other words, exercise should be a front-line therapy that can help bolster physical and mental health. It lowers the odds of conditions like gestational diabetes and preeclampsia (extremely high blood pressure) without increasing the likelihood of miscarriage, low birth weight, preterm birth or labor, or other birth complications. Currently, expectant people are advised to get at least 150 minutes of moderate-intensity activity each week throughout pregnancy if they don't have any medical or obstetric complications. Generally, people are advised to do what feels good, ideally some combination of aerobic and resistance training activities that don't pose a risk of falling or trauma to the body (like horseback riding or downhill skiing).

There are, of course, a few caveats that are common sense, like not working out in excessive heat, avoiding activities that make you feel light-headed or nauseous, and stopping if you're bleeding or experience chest pain or contractions. While there's no need to cap heart rate at

140 beats per minute, monitoring exercise intensity by using the "talk test," to ensure you can carry on a conversation while working out, is a good idea.

While these new evidence-based guidelines are a step forward, researchers still don't have a good handle on the impact of more vigorous physical activity on maternal and fetal health. Most studies on pregnancy and exercise have evaluated the effects of moderately intense exercise, the equivalent of an easy jog, but activities like high-intensity interval training, CrossFit, and endurance sports have all become increasingly popular and part of people's regular fitness in recent years. "You have all these different activities. Women want to continue that while they're pregnant, and we don't have the evidence to be able to support them or not," Davenport says.

The lack of clarity and answers sent Kikkan Randall down an internet rabbit hole. The five-time Olympian and three-time FIS World Cup Overall Sprint Champion and her husband knew they wanted to start a family after the 2014 Winter Olympics, but she had questions. *Lots* of them. Would her intense training over many years impact her chances of getting pregnant? What activities were safe to perform and at what level of intensity? How soon could she return to sports after childbirth? What protocols should she follow? "I just started rabidly Google searching and trying to find out about how other athletes had done it," Randall told me. But it was hit or miss. She had a hard time finding information on the physiological changes during pregnancy and how that might affect her as an athlete. In most cases, she came across differing opinions on what was considered safe.

It wasn't until 2015 that the International Olympic Committee (IOC) turned to the topic of pregnant and postpartum athletes. Recognizing that athletes were competing longer and didn't want to wait to start a family, the IOC gathered sixteen experts in Lausanne, Switzerland, to craft a consensus statement. What was meant to be one paper turned into five, covering areas related to pregnancy, maternal and fetal outcomes, and postpartum return to exercise. Kari Bø, a professor of sports medicine at the Norwegian School of Sport Sciences, led the IOC effort. She told me that while it was satisfying to sort through the existing literature, she was frustrated that they couldn't offer concrete recommendations about how much, how long, and how intensely pregnant athletes can or should exercise, or even the best path to return to sports. "[The athletes] are calling. They are sending emails. They want to have answers and responses, and we can't say yes, it's this," she says. (Preliminary research published since the IOC meetings does suggest that elite athletes don't experience more complicated deliveries such as a longer labor, need for cesarean sections, or perineal tears.)

For Abby Bales, pelvic floor physical therapist and founder of Reform Physical Therapy, the question isn't whether expectant people should exercise. It's how we can think about the possible ramifications of exercise on the pregnant body and implement strategies to combat and rehabilitate them, particularly when it comes to the pelvis.

To accommodate a baby, the pelvis tips forward, rotates externally, and gets wider, which warps the tension and relationship between the pelvis and the muscles and connective tissues attached to it. While these are all normal adaptations, muscles don't necessarily respond the way they're supposed to. Coupled with high-volume or high-intensity repetitive exercise and the additional downward pressure these activities place on the pelvic floor, there's a real prospect of layering risk factors for pelvic floor and musculoskeletal dysfunction on top of one another. Despite knowing that these changes will occur—and having a specific timeline for *when* they will occur—we still don't do enough to help people make their bodies more resilient for both pregnancy and the postpartum period.

———

While prenatal exercise recommendations are just starting to come around to more evidence-based standards, Margie Davenport describes the current guidelines for postpartum return to physical activity as "poor and weak." For more serious and elite athletes returning to competition, they're practically nonexistent.

The lack of attention reflects the long-standing assumption that the body naturally heals with little outside assistance and that it is normal for athletes—and women in general—to return quickly to form after giving birth. It's a narrative celebrated in the media with stories of "badass moms" achieving lofty professional milestones post-baby. British marathoner Paula Radcliffe started running twelve days after giving birth to her daughter, Isla, and went on to win the 2007 New York City Marathon ten months postpartum. Swimmer Dana Vollmer won three medals at the 2016 Rio Olympics seventeen months after her son was born, becoming USA Swimming's first mother to win a gold. Former WNBA player and basketball legend Sheryl Swoopes was back on the court six weeks after her son was born in 1997 and went on to help the Houston Comets win the WNBA championship that year.

American distance runner Kara Goucher also seemed to have a fairy-tale return to competitive life. A week after giving birth to her son, Colt, in September 2010, Goucher went for her first run. She was officially back at practice three weeks postpartum, and by December she was running more than 100 miles a week. Seven months after giving birth, she finished fifth in the women's race at the 2011 Boston Marathon (and set a personal record). Running publications ran stories on how to get your body back after baby and featured pictures of Goucher racing.

But Goucher told me she underestimated the immense changes her body went through during pregnancy and the strain of labor, which had left her feeling beat up. Her sacrum was "hot" for a few days after delivery and she thought she had cracked it. While her sacrum hadn't broken, her hip constantly hurt. At the time, no one was guiding Goucher. No one understood that she still wore what she described as "basically a diaper" for months because she was still bleeding or that, for six or seven months, she sat on a doughnut while driving because she still hadn't healed from childbirth. Ten months postpartum, she was diagnosed with a femoral head stress fracture.

Goucher says she has very few regrets in life, but she does regret rushing back to high-level training. "I think if I had just walked-jogged here and there, it would have been totally fine. But it was that getting back into intensity, doing an actual workout where I'm warmed up and spiked three weeks after giving birth to my son, that was a huge mistake," she says. Goucher was also surrounded by a coaching staff that was all men and people whom she described as "super hardcore." She says, "The few times I brought it up, I was told I need to toughen up and get back to my old self." Under immense pressure from her sponsor and coach at the time, she felt she had no choice but to push, push, push. The accumulation of trauma led to damage in her pelvic region. Chronic injuries ping-ponged back and forth between her femur, sacrum, and pelvis for the rest of her career.

The media, however, didn't get that story. Goucher wanted to be taken seriously as an athlete, which meant brushing aside any perceived sign of weakness, particularly any reference to her as a mother. In interviews, she purposefully avoided questions about pregnancy and parenthood. Looking back, she cringes at the message she likely sent to other women: I did this, so you can too. In a culture where women are judged by how quickly they get their pre-baby body back, you're perceived as lazy when you don't bounce back right away. Or maybe you just didn't want it badly enough. But Goucher says we should be asking, "How can we help you heal in the best way possible and the most complete way possible?"

The reality is that the physiological, hormonal, and anatomical changes that accompany pregnancy don't disappear immediately after birth. Since I kept up my exercise routine during pregnancy, I thought I would be back to light cardio at two weeks postpartum and running at six weeks. Instead, time that I thought I would spend building back my fitness was spent breastfeeding, pumping, or sitting on a sitz bath. The entire lower half of my body felt uncoordinated and unrecognizable. I had no idea what was happening or what was normal—I only knew this wasn't what I was told to expect. It wasn't necessarily that I wanted my pre-baby body back, although I'd be lying if I didn't acknowledge feeling the pressure around me. It was that I didn't feel like "me" and I had no idea how to get back there.

It can take a year or more for the body to return to "normal," depending on a person's birth experience, recovery, nutrition, and breastfeeding status. The muscles, connective tissues, and nerves of the pelvic floor bear the brunt of the trauma from pregnancy and childbirth as they stretch to great lengths to accommodate the additional weight of the fetus. After delivery, the region is like an overstretched elastic waistband that's lost its bounce. Approximately 80 percent of people also experience a vaginal tear during the birth of their first child, and for 40 to 50 percent of them, this involves the muscles around the pelvic floor.

Professional rock climber Beth Rodden told *Outside* in 2018 that in "the first few weeks after birth, I couldn't walk around without extreme pain and pressure and feeling like my insides were going to fall out." Muscles can have trouble firing and syncing with the rest of the core. One cough, sneeze, or even a funny step off the curb and the pelvic floor may not respond fast enough to avoid urine leakage. Without proper rehabilitation, damage to this area can lead to pelvic floor dysfunction. Up to 30 percent of people who have given birth for the first time will complain of urinary incontinence and 20 percent will suffer from fecal incontinence in the year after the birth of their first child. As many as half will experience pelvic organ prolapse.

The inconvenience of leaking, vaginal heaviness, and even pain with sex are assumed to come with the territory when you sign up for motherhood. "[People] are told it's normal. They feel like if they are bringing it up that they are somehow not coping with it well or that they're being difficult or needy and that is not the case," says Rita Deering, an assistant professor who specializes in pregnancy and postpartum physical therapy. These symptoms have major quality of life implications, and for active women and athletes they can have ramifications for exercise and performance. If you're worried about leaking, you may compensate in ways that could lead to injuries or chronic issues down the line. "If someone's holding back ten percent because they're afraid of leaking if they push too far, that's going to affect their performance and freedom to move," says UK-based advanced physiotherapy practitioner in pelvic health Gráinne Donnelly. And for elite athletes, that could mean the difference between standing on the podium or not.

Along with the pelvic floor, pregnancy also disrupts the body's core. As the baby grows, the connective tissue between the abdominal muscles widens and separates. Usually, the muscles come back together after childbirth, but for roughly one-third of people, the muscles remain separated, a condition called "diastasis recti." Aesthetically, it makes the stomach looks like it's bulging, but it's more than a cosmetic issue. When the connective tissue is stretched thin, it doesn't transfer force generated by the abdominal muscles effectively. That means postpartum people may need to drive their muscles more to produce the same amount of force as someone who has never carried a pregnancy, and they can tire faster. Women who have given birth (both vaginally and by C-section) also have weaker trunk muscles compared to women who have never been pregnant.

These muscles play an important role in protecting and stabilizing the spine and pelvis. "The impairments in strength and fatiguability can absolutely increase injury risk," Deering says. Olympic 800-meter runner and mom of three Alysia Montaño has openly discussed her experience with diastasis recti. While she has rehabilitated her abdominal wall, she says her connective tissue was "damaged beyond healing." In November 2020, she opted for surgery to repair her abdominal separation so she could function and perform at the level she needs to as a high-performance athlete.

Until the menstrual cycle resumes, Abby Bales says the body's hormonal environment is "pseudo-menopausal." As anyone who has given birth knows, it can be challenging to maintain a regular eating schedule and keep up with daily energy needs, especially during the early postpartum months and without caregiving help. It's easy to fall into an energy debt, increasing the risk for RED-S.

While exercise, even intense workouts, doesn't seem to impair the quality or quantity of breast milk production, the ripple effects from nursing—sleep deprivation, engorged (and painful) breasts, sporadic eating patterns, and fluctuating hormones—can affect how one feels day-to-day and during training. Most of what we know about the relationship between exercise and breastfeeding is based on anecdotal evidence shared behind the scenes between past and present nursing parents rather than scientific studies. And nursing people, even elite athletes, often scramble to crowdsource information wherever they can find it.

When Aliphine Tuliamuk was in the thick of her ramp-up for the Tokyo Olympics, she was breastfeeding her daughter, Zoe, who was about six months old at the time. She wondered if hard training and nursing were compatible, writing on Instagram, "Somedays I show up to workouts feeling low energy but workouts go great, I do wonder though if this is sustainable, I am doing my best to fuel properly but I also recognize that I could be playing with fire." She wanted to base her decision on research but couldn't find much.

She gathered information from other high-performing athletes on social media, most of whom were able to handle training while breastfeeding with some accommodations—pumping or nursing before their workout, wearing a good supportive sports bra, and eating and drinking enough. Weaning their babies off the breast had a greater impact on how they felt physically, mentally, and emotionally—it was a hormonal rollercoaster that left them feeling sluggish and off.

Still, these individual experiences represent case studies of one and may not be applicable to every body or every situation. Yet in the absence of other guidelines, these stories morph into the de facto playbook for postpartum people returning to physical activity.

D espite the monumental shifts in anatomy, physiology, and hormones during and after having a child, there's no comprehensive protocol for postpartum recovery and return to sports. In the United States, people customarily see their obstetrician once after childbirth, a routine checkup six weeks after delivery. If there aren't any red flags, they're told they can resume physical activity with no further guidance or education.

But the six-week clearance for exercise is an arbitrary time point, not a definitive milestone. "Six weeks is the demarcation for tissue healing," Abby Bales says. "That clearance is really a clearance of infections and post-delivery complications. It's not a clearance for activity." At that point, not all bodies will be ready to handle the load and impact of activities like running or high-intensity fitness classes. It can take four to six months for the largest pelvic floor muscle, and the connective tissue and nerves associated with it, to heal. For someone who has had a cesarean section, it may be six to seven months before the abdominal fascia returns to near-full strength. Only 31 percent of postpartum soldiers, the closest surrogate in the research literature for active women and athletes, equaled or bettered their pre-pregnancy fitness scores a year after giving birth.

But no one tells you how long it takes for your body to heal or what it really means to rehabilitate your body. No one really pays attention. When elite trail and ultrarunner Amy Leedham asked her doctor to examine her pelvic floor during her postpartum visit, she remembers the doctor telling her, "I don't really do that." Instead, women are merely told to rest and do some Kegels, despite one in four women performing the classic pelvic floor exercise incorrectly. Gráinne Donnelly told me that most people haven't been taught how to contract or relax their pelvic floor muscles, nor do they have a clear picture in their minds of where the muscles are and what they do. It's not like the biceps or quadriceps, muscles you see every day and know how to move. "How do we expect women to read a piece of paper with one simple instruction and get that right? It's just an impossible task," she says.

What's more, most obstetricians don't routinely refer women to pelvic floor physical therapists, specialists who can provide assistance and guidance. "Patients don't know we exist," Bales says. "They don't know that there's a process by which you can rehabilitate afterward because no one's told them." It was happenstance that Kikkan Randall realized she needed to pay attention to her pelvic floor. About a month after the birth of her son, she stood up from a chair and tweaked her back. Her friend, a physical therapist who happened to be studying the postpartum period, told Randall she wasn't using her core muscles and, as a result, the stress from big movements (like standing from a seated position) was channeled directly to her back and hips. "I was someone who was even proactively asking questions and looking for research, and I still didn't clue into it until I hurt my back," she says.

It's a standard of care that wouldn't hold up for an injury in any other part of the body. If you have a grade 2 (or moderate) injury to the hamstring, similar to a second-degree tear of the pelvic floor musculature that occurs in roughly 40 percent of births, a sports medicine or orthopedic physician prescribes time off along with a referral to physical therapy. The physical therapist then puts together a step-by-step plan for the next two to three months designed to help the muscle get stronger and progressively tolerate more load. They demonstrate the exercises, then watch as you perform them, fine-tuning your movements as needed. This rarely exists for postpartum injury and trauma. Unlike a sports injury, which is unexpected, we can proactively prepare and react to the changes that accompany pregnancy and the postpartum period, and have nine to ten months to do so.

Bodies need time and assistance to come back together after giving birth and to settle into a new postpartum normal. There's growing research demonstrating that conditions like pelvic floor dysfunction and diastasis recti aren't foregone conclusions; they can be rehabilitated. "You need to spend some time learning how to reactivate all the muscles, coordinate them, and then restrengthen and rebuild core control," says Celeste Goodson, a medical exercise specialist and founder of ReCORE Fitness, who has worked with a number of pro athletes pre- and post-baby.

Bales advocates for postnatal screening for everyone, regardless of mode of delivery, and a referral to a pelvic floor physical therapist at any sign of dysfunction. When patients come to her, she assesses their ability to voluntarily contract and relax the pelvic floor muscles and helps them reestablish the brain-muscle connection. Often, she finds that people will bear down when asked to engage their pelvic floor rather than drawing the muscles up and in, a sensation Bales describes like squeezing a pencil inside the vaginal canal and lifting it up toward the belly button. To cue them, she'll suggest that patients imagine they're upside down in yoga class and don't want to pass gas. "Their rectum is surrounded by the same muscle that surrounds their vagina, and so when you get a contraction there, you say, 'Yeah, that's what that feels like,'" she says. Treatment is then tailored to each patient and can include manual therapy, biofeedback, and exercises to improve the function, strength, and stability of the core and pelvis, like pelvic tilts, clamshells, bridges, and wall sits.

Kikkan Randall's physical therapist focused on her breathing. To help her reconnect with her diaphragm, Randall practiced blowing up balloons as her therapist observed. She says strengthening her "internal frame" paid off. "It's those little muscles that hold you together when you're falling apart late in a race. That was a cool upside that I never would have discovered without pregnancy," she told me. "Regardless of whether you're an athlete or not, every woman should be getting guidance."

You can also retrain the abdominal muscles and decrease the gap in the abdominal wall (the inter-recti distance). Rita Deering led a study with recreational runners who ranged from seven weeks to two years postpartum. The runners started by learning how to activate the transversus abdominis, the deep layer of abdominal muscle, before progressing to strength training exercises like hip bridges, side plank on knees, bird dogs, and single-leg squats. In the lab, researchers used ultrasound so the women could see the deep abdominal muscles contracting, imprinting how the exercises were supposed to feel. After eight weeks, the group had reduced the amount of separation below the belly button.

Interestingly, Deering told me that in three women, the inter-recti distance got worse. These women were in the early postpartum phase when they enrolled in the study, between seven and twelve weeks after childbirth, and were already running at least 6 miles a week. By the end of the study, they were back to their pre-pregnancy mileage. To Deering, their experience suggests they may have returned to running before their bodies were ready.

Sometimes people need guardrails to keep them from overdoing it. In the later stages of pregnancy and the first eight weeks after childbirth, Aliphine Tuliamuk's pelvic floor physical therapist became her de facto coach. She texted Tuliamuk before her six-week postpartum appointment and told her that while the doctor would likely clear the marathoner to run, she wanted her to wait two more weeks. Without that guidance, Tuliamuk told me that if her doctor said it was okay to run, she would have. "But I was in no place to start running at that point," she says. The two extra weeks made an important difference—she felt her body could truly handle the increased load without falling apart. "It was insane how much recovery I had from week six to the end of week eight," she says.

On Mother's Day 2019, *The New York Times* published a video op-ed from pro runner Alysia Montaño. It was a play on Nike's "Dream Crazier" ad campaign, a video that featured snippets of women athletes pursuing audacious goals. Montaño's video, on the other hand, showed how pro sports and motherhood were incompatible and, frankly, a crazy idea. Montaño and others detailed how athletes can lose pay if they become pregnant, and how companies like Nike and ASICS could pause or reduce contracts if athletes didn't meet certain performance thresholds. There was no exception for pregnancy or postpartum recovery, even though an athlete's body is their livelihood and that body has endured significant physiological changes and trauma. Montaño said, "The sports industry allows men to have a full career and when a woman decides to have a baby, it pushes women out at their prime."

Historically, there has been little incentive to invest in pregnant and postpartum athletes or give them the opportunity to recover. "In this world we're unwilling to give someone time, and after having a child that's what you need," Allyson Felix told *The New York Times*. When

Felix's contract with Nike neared expiration in 2017, Nike initially proposed a 70 percent pay cut to the most decorated American track and
field Olympian in history. Felix had learned she was pregnant in the
midst of negotiations. Afraid Nike would rescind their offer, she hid her
pregnancy. Felix worked out early in the morning, wore baggy clothes,
and kept her baby shower small to prevent word from getting out. The
negotiations dragged on after the birth of her daughter, Camryn, in
November 2018, and Felix felt pressure to return to form quickly, even
though she experienced severe preeclampsia, threatening her and her
daughter's lives, and underwent an emergency C-section at thirty-two
weeks. She asked Nike to guarantee that she wouldn't be penalized
if she didn't perform her best after giving birth, and she says they
declined.

Pro snowboarder Kimmy Fasani was similarly terrified to tell her
sponsor Burton that she was pregnant. At the time, her contract was
also up for renewal and didn't include a maternity clause, but Donna
Carpenter, who helped build Burton with her late husband, Jake, was
all ears. Inherently, the two parties trusted each other, and they figured
out a reasonable timeline that gave Fasani breathing room after her
son's birth in 2018. "We were both in a gray area," Fasani told me, but
she knew Burton stood behind her because the company saw her value
outside of just being an athlete. Instead of rushing back according to an
arbitrary timeline, she knew she could go to her sponsor and tell them
she wasn't ready to return yet.

Following Montano's op-ed video, Burton changed all of its contracts with women athletes to include the same maternity support and
protections Fasani enjoyed—no reductions or suspensions of pay for
six months, no contract termination due to a pregnancy or maternity-
related reason, accommodations for contractual obligations, and support for a companion to travel with the athlete to required appearances

while they're breastfeeding. Several companies revised their contracts to include maternity policies too, including Brooks, Altra, Nuun, and even Nike.

When sponsors recognize athletes as more than just a time on the clock, a judge's score, or podium finish, it fosters a working environment that prioritizes women's safe return to and longevity in sports. Distance runner Stephanie Bruce, like Fasani, had the full support of her sponsors during her two pregnancies. In fact, Bruce was seven months postpartum and pregnant with her second son when she signed with the Northern Arizona Elite (NAZ Elite) team in 2015, sponsored by Hoka. Being pregnant and a new mom isn't exactly ideal when negotiating with a new team and sponsor. There was a giant question mark about her future, and Bruce essentially asked them to make a huge bet on her. Hoka and Oiselle (Bruce's other main sponsor at the time) went all in and didn't put any stipulations on her return to competition.

Bruce needed the flexibility and support. With her first pregnancy, she had an episiotomy, where doctors make an incision to widen the vaginal opening, and a grade 4 tear in her pelvic floor muscles that led to incontinence and pelvic floor dysfunction. On her first jog at eight or nine weeks postpartum, she made it three minutes before she told me she "pretty much shit my pants." While she thought there was an outside chance she'd qualify for the 2016 Olympic marathon team, her pelvis kept giving her trouble. If she had a different sponsor, she might have felt pressure to go for it, but she knew it wasn't worth sacrificing her body for this one race and she had the option to put her health first. "That's when my perspective changed. I just started to have more patience in my timeline. I gave myself grace," she told me.

Bruce returned to competition after two years off, and she ran better than ever. She won her first national title in 2018 at the U.S. 10K championships at the Peachtree Road Race in Atlanta, Georgia. Between

2019 and 2021, she notched personal bests in the 5K, 10K, half-marathon, and marathon. Bruce's story is a win-win for her, her sponsors, and sports. Kara Goucher told me, "She's been a great example of looking at what an athlete's career can be when we look at them as a lifelong athlete versus the short term. She's going to be providing stories and races and results for years to come."

Recently, better guidance has begun to emerge for postpartum people. In 2018, ACOG proposed "a new paradigm for postpartum care." Instead of a single visit at six weeks after giving birth, they now recommend that people receive ongoing care during the first twelve weeks postpartum, with the first visit scheduled three weeks after delivery. They call this period the "fourth trimester," a recognition that people need ongoing advice and support to optimize their physical, emotional, and mental health.

In 2019, Gráinne Donnelly, together with UK-based physiotherapists Tom Goom and Emma Brockwell, released *Returning to Running Postnatal—Guidelines for Medical, Health and Fitness Professionals Managing This Population*, one of the first evidence-based guidelines of its kind, based on current research and the authors' years of clinical experience. They focused on running because it's a popular and accessible form of exercise, and it's often the first step back to other high-intensity activities and sports.

The hope is that this step-by-step framework stems the mad rush back to exercise after childbirth, particularly to high-impact activities that can set women up for long-term problems. If there are no signs of dysfunction, they recommend a graded return-to-run program where the first six to twelve weeks focus on the basics—learning to reengage the core and pelvic floor muscles and restoring the synergistic relationship between them. Then you can begin to work progressively on strength (squats, lunges, bridges) and cardiovascular fitness, starting with low-impact activities like walking, so that, when you go back to higher-impact activities, your body is ready. In general, they say you shouldn't ease back into running and more sports-specific activities until three months after delivery.

Still, the current system of postpartum care remains fragmented and doesn't fully consider the multifaceted needs of pregnant and postpartum people. Physical health, pelvic health, emotional well-being, nutrition, sleep, breastfeeding, and exercise all need to be part of the equation. In particular, pelvic floor physical therapy should be included as part of the routine pregnancy and postpartum care model. If therapists worked in conjunction with obstetricians and on labor and delivery floors, they could potentially prevent troublesome issues—like perineal tears and pelvic floor muscle trauma—before they start, rather than react to dysfunction once it's already set in. We need to continue to build better evidence-based protocols and a comprehensive, progressive framework to steer exercising women and athletes back to sports. We also need more research on heavy lifting, high-intensity exercise, and longer duration of exercise during pregnancy and the postpartum period, especially pertaining to elite athletes.

Sports governing bodies, federations, and sponsors need to do a better job of helping their athletes too. Pregnant and postpartum athletes shouldn't have to rely on Google or anecdotal information to answer their questions. When Elana Meyers Taylor started her family, her coaches (who were men) tried to help, but it was hard to explain what she was going through, whether it was menstrual cramps or trying to make weight while breastfeeding. Their advice was based on their wives' experiences, which weren't applicable to the five-time Olympic medalist and two-time world champion in bobsled.

Meyers Taylor and Kikkan Randall are part of a group of Olympians pushing the U.S. Olympic and Paralympic Committee to step up and support women athletes, particularly when it comes to family planning, pregnancy, and the postpartum period. Meyers Taylor dubbed the group "the Mom Avengers." Randall says, "The U.S. Olympic and Paralympic Committee is world-renowned for their sports science research. If we can get this to be one of their priorities, then there would be a lot of great resources at our disposal."

As the conversation around pregnant and postpartum bodies is evolving, people are recognizing that having children and being physically active don't have to be mutually exclusive—women can have a family and a fully active life, even at the highest levels of sports. And women athletes are proving they are worthy of investment and support.

In June 2021, Allyson Felix vied for her fifth Olympic team and first since becoming a mother. On finals day for the women's 400 meters at the U.S. Olympic Track and Field Trials, Felix lined up in lane eight. She was far behind on the final turn, but heading into the final stretch, it was as if muscle memory kicked in. All those years of winning medals and championships powered Felix to the finish line. She placed second while, two lanes over in lane six, Quanera Hayes—mom of a two-and-a-half-year-old son—won. Both athletes secured their spots for the Tokyo Olympics. Afterward, Felix's daughter, Camryn, and Hayes's son, Demetrius, joined their moms on the track to celebrate. The two kids hugged, and Felix told them, "Guys, we're going to Tokyo!"

Sitting on the track with Camryn between her legs, still catching her breath, Felix said, "It has been a fight to get here"—fighting in the neonatal intensive care unit, fighting for her life, fighting to get back on the track, and fighting for sponsorship. In the post-race press conference, Felix said, "Society tells us a lot of times you have a child and your best moments are behind you, but that's absolutely not the case. I am a representation of that. Quanera is. There are so many women across industries who are out here doing it and getting it done."

And Felix got it done. At the Tokyo Olympics, she won a bronze in the 400 meters and a gold in the 4×400-meter relay. With eleven career Olympic medals, she became the most-decorated U.S. track and field athlete of all time, woman or man. Felix is not only rewriting the history books and reimagining what it means to be a mother-athlete. She's showing that if we just leave ourselves open to opportunity, active lives don't have an expiration date.

11.

The Change

⋅∘∙●●●∙∘⋅

Humans are suckers for progress. We like to mark growth, whether it's ticking off the weeks of a training plan or plotting our kids' height on the wall. We take special pride in milestones that represent a shift from one stage of life to the next. Traditions like red egg and ginger parties in Chinese culture celebrate new additions to the family. Bat mitzvahs and quinceañeras mark the passage into adulthood. Elaborate ceremonies declare commitments to life partners.

But when it comes to menopause, the transition to mature adulthood is relegated to obscurity. There's no celebration or rite of passage. There's no comprehensive library of "what to expect" resources. There's no educational classes or networks to lean on. There aren't even clear signs—like budding breasts, pubic hair, or a baby bump—to confirm that you've entered this new phase until you're in the thick of it. The most common refrain I've heard from people who've entered menopause is, "Why didn't anyone tell me about this?"

For centuries, women were primarily valued for their reproductive capacity, so the cessation of one's menstrual cycle and fertility in midlife was seen as a troubling development. The sentiment is reflected in the ways menopause is described—as ovarian or hormonal deficiency, failure, and exhaustion—reinforcing the cultural perception that middle-aged women themselves are past their prime and no longer relevant or useful to society. They're assumed to be in the sunset years of their lives like the women on *The Golden Girls*, one step away from the grave. It's no wonder that people aren't eager to draw attention to their new status.

Instead, menopause is shrouded in myths and polite euphemisms like "the change," and people largely navigate the transition isolated and alone. Steph Creaturo, a yoga instructor and run coach based in Brooklyn, New York, described her experience with the menopause transition like being sideswiped. Then she paused, took a breath, and clarified what she meant. She told me, "It felt like I was washed over by a tsunami out of nowhere." But it wasn't a normal tsunami. It was a tsunami laced with fire, monsters, dragons, and spikes like you'd see in a video game.

When Creaturo was forty-seven, she noticed her period "getting weird" before everything "fell off the cliff." Her body no longer felt like her own, which was disorienting. Whenever she wanted to talk about what she was going through or commiserate with others, people gave her the cold shoulder. No one believed her world could be so askew. When she brought it up on group runs, people changed the conversation or gave her the side-eye and asked her to stop making such a big deal out of nothing. Physical therapists assumed that her aches and pains were due to running. Her doctors didn't raise any red flags. It couldn't be menopause; she was too young for that.

Eventually, Creaturo chalked up her symptoms to side effects of life and age—her roles as mother, wife, and New York City business owner converging into a giant stress vortex. What other explanation was there? She figured she just had to suck it up and deal with it. She quit her running team and didn't talk about her symptoms, uncharacteristically hiding her distress from everyone, including close friends and family. Creaturo's the type of friend who holds your water bottle on a run and has it right where you need it at mile 20 of a marathon. She's always down to have a real conversation, and it was the first time in her life she wasn't 100 percent honest. She became depressed in a way she'd never experienced before.

Yet, for anyone with ovaries, menopause is simply another life stage, and people spend between one-third to one-half of their life postmenopause. It's the counterpart to puberty, when the ovaries wake up and begin their cyclical rhythm of ovulation and hormone production. Around the midforties, the ovaries start to run out of viable follicles (the structure that contains the maturing egg and releases it during ovulation), which throws off the production of estrogen and progesterone and begins the menopause transition. Progesterone drops first, and without it to serve as a counterbalance to estrogen, estrogen begins to fluctuate wildly. The changing hormone levels look like a toddler's scribble that ends in an abrupt drop-off rather than a neat, regular rise and fall. When there are no more viable follicles left, ovulation stops, and without ovulation, there's no menstrual cycle or regular production of hormones. (Fat tissue, skin, kidneys, and the brain continue to produce some estrogen, but the hormone generally acts locally on those specific tissues.)

Technically, menopause is defined as a full year without menstruation, which, on average, occurs around age fifty-one. But it's impossible to know when a person's last period is really their last period. There's no screening test to definitively say someone is premenopausal on one side and postmenopausal on the other side. Doctors make the diagnosis based on a person's medical history and symptoms. For a small number of people, periods can stop before the age of forty, a condition called "primary ovarian insufficiency." For others, certain medications (most often for cancer treatment) or surgery (such as removing the ovaries) can bring about menopause.

While menopause is marked by the last menstrual period, it's so much more than that. Oscillating hormones, and the abrupt withdrawal of those hormones, can induce a long list of symptoms. Vaginal dryness, depression, anxiety, joint pain, heart palpitations, and "menopot"—a term for the potbelly that can take shape in midlife—are all common. Creaturo logged 332 straight days of hot flashes and night sweats before she stopped counting. She woke up drenched in sweat and had to wash her sheets every day. Her hair fell out. She was bloated and constipated. Her metabolism fluctuated along with her weight. Physically, she was exhausted, like someone let the air out of her tires. A standard, easy run rendered her breathless. "I had no more pep in my step. I had no more spring," she told me. Mentally, Creaturo's cognitive abilities were so clouded that at times she felt possessed. Bouts of brain fog in the middle of teaching yoga left her grasping for the names of poses she has taught for more than twenty years.

Symptoms tend to be worse during perimenopause—the years leading up to the final menstrual period—before gradually easing after menopause, but the lived experience of the menopause transition can differ greatly from person to person. Often, there's no rhyme or reason to the symptoms or any patterns to track, contributing to the confusion and fear about how the body is changing. The timeline is nebulous too, lasting up to ten years before the final menstrual period and another five to ten years after. Data from the Study of Women's Health Across the Nation seems to suggest that the social factors associated with race may play a role in the menopause transition (although race alone wasn't a statistically significant variable). Black and Chinese women, and Latinas, on average, enter the menopause transition earlier than white women and can have longer transition periods. Symptoms—both the severity and how long they last—can also vary depending on whether you exercise, take birth control or other medications, smoke, or drink, as well as many other medical and lifestyle considerations.

Creaturo was postmenopausal by age forty-nine. Looking back, she believes there were clear signs that her hormone levels were dropping precipitously and chaotically. Except no one knew she was experiencing menopause. No one could help her because they didn't have a road map, or even a name, for what she was going through. She'd suffered by herself, and with zero support.

Stories like Creaturo's terrify me. As a cisgender woman in my mid-forties, I know menopause is a fuzzy date on the horizon, fast approaching. Still, I don't know what it *really* means, and I have a hunch that the stories I've heard only tell half the story. It feels eerily familiar to my postpartum experience, when friends and family put on a smile and raved about motherhood, glossing over the not-so-glamorous parts. It wasn't until I officially joined their ranks that they spilled their secrets and gripes. With menopause, it seems like there's no way to prepare for this new stage of life—physically, mentally, or emotionally—because no one talks about it.

Menopause has long had a public relations problem, thanks to a deleterious combination of sexism and ageism. Historically, it was thought to foretell the deterioration of physical health. One of the first mentions of menopause appeared in the 1710 paper "On the End of Menstruation as the Time for the Beginning of Various Diseases," a title that reflected the prevailing belief that menopause represented fading vitality and health. University of Georgia professor of history Susan Mattern wrote in *The New York Times* that physicians in Europe believed women accumulated blood and toxins in the womb when they didn't have a regular period and blamed a multitude of medical

ailments, from cancer to epilepsy to "hysteria," on dysfunctional reproductive organs. Victorian doctors even thought that postmenopausal
women would suddenly sprout scales on their breasts.

In 1821, French physician Charles de Gardanne coined the term
"menopause," combining the Greek words *men* (meaning "month") and
pausis (meaning "stop"). Once it was defined as a medical condition,
menopause became a disease to be cured. In the late nineteenth century, scientists discovered substances associated with the ovaries (later
identified as hormones) that seemed to play a role in the midlife transition. Several studies in Europe found that when women who were either
menopausal or had their ovaries removed were injected with ovarian
tissues from cows, their menopause-related symptoms improved.

Pharmaceutical companies jumped on the bandwagon and wanted
to commercialize this remedy. In the 1890s, Merck & Co. developed
Ovariin from powdered cow ovaries. From there, a whole suite of hormone therapies was designed to replace estrogen and preserve a woman's youthfulness, femininity, and sprightliness. Pills, injections, and
patches grew in popularity throughout the twentieth century, and by
1975, estrogen was the fifth most prescribed medication in the United
States. Women and their healthcare providers believed these treatments could solve the problem of menopause.

In the 1990s, the National Institutes of Health (NIH) launched a
clinical trial to evaluate menopause hormone therapy (MHT, also known
as hormone replacement therapy) and determine if it could improve
bone density and cardiovascular health. However, a July 2002 study in
The Journal of the American Medical Association (JAMA) found that
women who took MHT had an increased risk for breast cancer, heart
attack, and stroke, and researchers halted the study three years early.
News of the report blew up in the press, demonizing MHT. Wulf Utian,
then the executive director of the North American Menopause Society
(NAMS) told *The New York Times*, "This is the biggest bombshell that

ever hit in my 30-something years in the menopause area." The subsequent media storm reinforced the negative stereotype surrounding menopause and the notion that there were no viable treatments for it. Unsurprisingly, MHT use dropped 45 percent in the United States as the medical community panicked.

Years later, the *JAMA* study was widely criticized for not clarifying the clinical implications of its findings. It turns out that when it comes to the efficacy of MHT, a person's age and distance from menopause matters. For women who were younger than sixty or within ten years of their final menstrual period, hormone therapy had powerful benefits. Long-term follow-up studies showed that people on MHT were not at a greater risk of dying, and those taking estrogen therapy alone had a significantly lower risk of breast cancer diagnosis. But the damage had been done. Doctors are reluctant to prescribe MHT, and women are hesitant to take it, even though NAMS, one of the leading medical societies for women's health, believes it can be a first line of defense for the right women, at the right time, with the right indications.

These events had a chilling effect on the medical field and the study of menopause, leaving it in a no-woman's land of healthcare. There is no team captain to champion the cause of middle-aged and older people with ovaries. Obstetrician-gynecologists (ob-gyns)—those who typically address women's health needs and whom women consult for much of their adult life—largely focus on a person's childbearing years and prenatal care, which is no longer relevant during menopause. In the current medical environment, ob-gyns have also become increasingly hyperfocused on specialties like infertility, obstetric care, or maternal-fetal medicine, which means middle-aged people have nowhere to turn.

"There's a knowledge gap when it comes to providing comprehensive care for midlife women" and understanding the physiological changes to cardiovascular health, bone health, and mental health that accompany that phase of life, says Anna Camille Moreno, medical director for the Midlife Women's Health and Menopausal Medicine program at the University of Utah Health and a NAMS-certified menopause provider. According to a 2019 survey of family medicine, internal medicine, and ob-gyn residents across twenty residency programs in the United States, less than 7 percent felt they were prepared to help someone manage the menopause transition. When subjects like menopause aren't taught as part of the routine medical school or residency curriculum, it's left to residents and physicians themselves to stay up to date on the latest research and seek out specialty training.

In practical terms, the lack of rigorous formal training means medical professionals don't always connect the dots between the wide range of symptoms that emerge during this stage of life, even though an estimated 1.3 million menstruating people in the United States enter menopause each year. Some women report that doctors shrug their shoulders and attribute their symptoms to stress and aging. Other physicians just don't know what's going on, despite their best efforts.

Amanda Thebe was forty-two when her normal boxing class left her feeling unusually fatigued and in bed for two days. She had a bad case of vertigo and had to crawl on all fours to the bathroom. Thebe, a personal trainer and nutrition coach, figured it was a weird virus, but it happened again and again. Sometimes she lost feeling in her hands and face. Sometimes she had double vision. Each time, she was extremely tired.

Over the course of two years, Thebe saw multiple specialists— neurologists; ear, nose, and throat doctors; vestibular rehabilitation therapists—and underwent MRIs and CT scans to get to the bottom of her crippling symptoms. Doctors saw that she was unwell but couldn't figure out what was wrong. All her tests were inconclusive. She thought, "Maybe this is just who I am now?"—a fraction of her normally exuberant, extroverted self. She fell into a deep depression and rarely left the house. Like most people, Thebe didn't recognize the signs of the menopause transition on her own.

Moreno told me that by the time women find their way to her clinic, they've already exhausted multiple avenues trying to figure out what's happening to their bodies and minds, and they're frustrated. They've done a fair amount of googling, sleuthing, and searching for second and third opinions before they land on a site like NAMS where they can find a certified menopause provider. Moreno says a woman's initial visit with her is often emotional. Some start crying even before Moreno has a chance to say hello. "They want to know what's going on. They don't want to be ignored. They want to be heard," she told me.

Menopause can be, without a doubt, a disorienting experience, especially for active and performance-driven people with ovaries in their forties, fifties, and sixties. There's a familiarity that's born from countless hours spent testing the body's limits and becoming intimately attuned to its every nook and cranny. Then, suddenly, the body takes on an unfamiliar shape and just *feels* different. Tried-and-true training and nutrition plans no longer do the trick. Not to mention, it can be harder to carve out dedicated time for exercise given competing demands on one's attention and time, particularly between work and family responsibilities like caring for children and aging parents. It's no wonder middle age is often seen as the off-ramp from an active life. Older people aren't expected to move and perform at the same level, whether it's in a fitness class, a recreational league, or an elite race. They're expected to slink off and let the new generation step up and set all the records.

Currently, there's a dearth of information on the physiological changes that accompany midlife and how those changes impact active and athletic people. Scientists aren't sure whether peri- and postmenopausal people need to approach exercise, training, nutrition, recovery, and injury prevention differently than those who are premenopausal because there hasn't been an incentive to investigate. Previous generations weren't necessarily invested in physical activity or interested in remaining active in sports through midlife and beyond to the same degree as people today.

That's not to say there wasn't any interest in studying the effect of exercise on middle-aged populations. In 1991, NIH's first woman director, Bernadine Healy, announced the Women's Health Initiative (WHI), an estimated $1 billion research project to study women's health, one of the biggest studies involving women in the United States. Over a fifteen-year period, WHI enrolled 68,000 postmenopausal women, between the ages of fifty and seventy-nine, to understand how measures— such as exercise, diet, smoking cessation, hormone therapy, and dietary supplements—might prevent diseases that become more common after menopause, like cardiovascular disease, osteoporosis, and cancer, which are a "major cause of death, disability, and frailty in older women of all races and socioeconomic backgrounds," according to the National Heart, Lung, and Blood Institute. Scientists and the medical community wanted to know what could be done to prevent, slow, or stop the progression of these chronic conditions and to reduce mortality. (The WHI included the MHT clinical trials previously mentioned.)

In concert with the 1993 NIH policy that required scientists to include women and minorities in clinical trials, WHI (and its funding) drove research priorities and what scientists chose to study. When scientists considered exercise, they focused on the public health benefits of physical activity and determining whether exercise was an effective way to improve long-term health. For instance, during the second half of the twentieth century, there was an uptick of scientific studies involving women and resistance training, from three articles published in the 1950s to 287 articles published in the 1990s. "More than half of [the studies in the 1990s were] really about exploring aging, osteoporosis, and the loss of muscle mass that comes with aging," professor and sports historian Jan Todd told me. Scientists weren't trying to understand the limits of a *woman's* strength, or how best to train women athletes and build muscle mass.

The lack of research leaves little guidance for peri- and postmenopausal people, especially those who want to improve their fitness or continue competing in sports. As fitness, strength, and performance gains slow or come to a halt in inexplicable ways, the natural inclination is to redouble your efforts in order to stay in shape or be race ready. That's the message I gathered from fitness magazines, Instagram, and sports media: age is just a number and there's nothing standing in the way of my athletic goals except hard work. But what I didn't realize was that I was fighting against the tide of oncoming changes and that there were consequences to forging ahead with the same old training, nutrition, and injury prevention strategies.

"You might try to push harder and longer to get rid of this new belly fat, but ultimately that backfires, because it puts you in a state of low-energy, high-stress cortisol cycling," Stacy Sims, one of the leading experts on female physiology, told *Outside* magazine. It could also lead to overtraining and overuse injuries. Or it may leave you so frustrated that you raise the white flag, back off your training, or give up on physical activity entirely.

The peri- and postmenopausal years don't need to signal a fade to black for someone's active pursuits. It's just another stage of life that requires adjustment and a new mindset. And that starts with understanding what's happening in the body and anticipating and working with the changes.

In recent years, researchers and scientists have started to piece together the exercise and menopause puzzle. They're starting to understand how aerobic fitness, body composition, and strength change with age and what that might mean for active people with ovaries. They've found that there are perfectly reasonable explanations for why fitness and performance dip during the middle years of life.

The physiological shifts that accompany this life stage are due to a combination of multiple factors—some that are simply related to age and affect everyone, and some that are specific to the hormonal changes that accompany the menopause transition. Aerobic fitness decreases with age, regardless of sex. That's why many sports use age grading—adjustments to reflect how someone would have performed in their athletic prime. For example, to qualify for the Boston Marathon, the pinnacle in the sport of long-distance running, runners must meet qualifying standards based on their age and sex. With each jump up in age bracket, the qualifying time increases by an increment of between five and fifteen minutes. The accommodations make sense when you consider that VO_2 max, the maximum amount of oxygen the body uses during exercise and an oft-used shorthand for aerobic capacity, starts to drop between 5 and 9 percent every ten years beginning around age thirty-five in *all* healthy humans. By age sixty, running performance plummets even faster, particularly for women.

Body composition and proportions morph in midlife too, as people gain weight and take on a softer, rounder shape. One reason is that people in menopause tend to use and burn considerably less energy than they did previously. They have a lower resting metabolic rate (menopausal people even expend one and a half times less energy while asleep compared to those who haven't yet transitioned to menopause), expend less energy during the day (one study found menopausal people burn 200 calories a day less than premenopausal people), and don't engage in as much physical activity as they used to (30 percent of women say they're less active during menopause). Plus, the body becomes more insulin resistant due to falling estrogen levels and it no longer processes carbohydrates efficiently, leading to an accumulation of fat around the midsection and wild blood sugar highs and lows. Fluctuating blood sugar can cause cycles of hunger and exhaustion and makes fueling workouts, recovery, and everyday life tricky.

In this stage of life, the accumulating fat is more likely to be visceral fat (the deep abdominal fat linked to an increased risk for cardiovascular disease and type 2 diabetes) rather than subcutaneous fat (the kind that's just underneath the skin). Prior to menopause, visceral fat constitutes 5 to 8 percent of total body fat, but after menopause it accounts for 15 to 20 percent. Even when young and older women athletes have similar total body fat percentages and lean muscle mass, the older athletes still have more visceral fat compared to younger athletes.

Competitive athletes aren't immune to the body's shape-shifting either. Selene Yeager, a former elite bike racer and coauthor of *Roar* with Stacy Sims, told me she was confident she'd sail through the transition because she was an athlete. But when she hit her late forties, she didn't recognize herself in the mirror. On top of night sweats and anxiety, she gained fat in places she never had before, and her muscles retreated. Yeager also couldn't produce as much power on her rides as she entered middle age. She felt slower and less peppy during workouts, like someone turned up Earth's gravitational pull. It's most likely the consequence of tanking hormones like estrogen and testosterone, which shift the balance toward muscle loss rather than muscle gain. The body now disassembles muscle protein faster than it creates it and doesn't respond to training and nutrition stimuli like it did during younger years. In short, it's harder to build and maintain lean muscle mass with age.

The changes in muscle integrity and structure affect all muscles, including those of the pelvic floor. Muscles may be slower to fire or don't work like they used to, leading to issues like stress or urge incontinence and pelvic organ prolapse, regardless of whether a person has birthed a baby or not. With dropping levels of estrogen, women may experience vaginal dryness and atrophy, symptoms that are troublesome on their own but can cause chafing and discomfort for cyclists, runners, rock climbers, and others. These symptoms, which affect approximately 60 percent of people in menopause, can create yet another barrier to physical activity in middle age that leads some to abandon exercise altogether.

Understanding and anticipating these inevitable changes creates an opportunity to work with your physiology and make the transition a little less turbulent. Steph Creaturo reminded herself, "I'm not my estrogen," even though she was acutely affected by the changes in her hormone levels. She says she's settled into the "adultness of menopause," determining what she needs to do in her athletic and daily life to support her body as it is now and to slay her menopause symptoms.

Creaturo recognizes that her younger self is in the rearview mirror. "If I keep chasing her, is that really fair to who I am now? I have no idea what this fifty-one-year-old is capable of. I have to give her a shot," she says.

The truth is that your "prime" active years don't have to coincide with your fertility. Menopause opens a new chapter of your active life, not the inevitable beginning of the end. You don't have to give up the high-intensity training, endurance sports, fitness classes, or other adventurous pursuits you love. In fact, maintaining a regular exercise habit can prepare the body as well as offset the symptoms that come with the menopause transition. When researchers followed women in Finland classified as in pre, early, or late menopause, they found that physical activity buffered them from some of the age- and menopausal-related slowdown. The more active people were during peri- and post-menopause, the better they performed on tests of muscle strength and power compared to their less active counterparts.

Still, fitness and performance don't follow linear trajectories. Like the adolescent transition, the body may become less efficient and less responsive to exercise in ways that can feel counterproductive to an active life. It takes time to ride out the jarring waves of hormonal ups and downs until the body settles into its new normal, and it takes a fresh perspective to reframe the possibilities of this phase of physical activity. A novel approach to exercise is necessary instead of stubbornly doing things the same way.

While we wait for more research on menopause and exercise science to emerge, it's even more important to listen to your body, especially if symptoms negatively impact your quality of life. "You really have to tune into how your body's feeling and know that sometimes pushing through the fatigue is actually the worst thing the body needs," says Amanda Thebe. When Thebe was in the thick of her debilitating perimenopause symptoms, she had to ask herself how she really felt—physically, mentally, energetically—rating herself on a scale from one to ten. She then determined what she was realistically capable of doing that day. She adopted what she called "fitness snacking," little bite-size chunks of movement, physical activity, and strength training that gave her the flexibility to work around her energy level while still doing something that made her feel better.

If her energy level was a one or two, she didn't do more than go for a walk. When she felt like a three or four, she'd do some mobility work or bodyweight exercises. When her energy reached the five-to-seven range, she'd go to the gym and aim for ten minutes of strength training. Often, once she started, she did more than she expected. Not only did it give her a win, but she started to feel better too. Slowly, she rebuilt baseline, foundational strength so that on days when she woke up and felt normal, she could tackle a full workout.

Thebe's experience points to the importance of recovery, especially since cortisol can rise unchecked during menopause, nudging the body into a constant state of high alert. Add in sleep troubles—which plague approximately 40 percent of perimenopausal people and up to 60 percent of postmenopausal people—and the body's recovery cycle is significantly disrupted. Steph Creaturo's night sweats left her chronically fatigued and dehydrated. That lack of good sleep and hydration created a self-perpetuating spiral. Workouts were hard because Creaturo was always tired, and because she couldn't get good rest, her body never fully recovered or adapted to training stress, stalling any and all progress. Now she works *with* instead of against her fractured sleep schedule. No more early morning yoga clients or runs and no more teaching evening classes either, unless she can teach from home via Zoom. She also consults with a NAMS-certified menopause practitioner and takes hormone therapy, which she claims saved her life during the menopause transition.

Experts say that when you're ready to train or get back to regular workouts, the goal is to use exercise and nutrition as a way to create a more stable internal environment, to spark a specific type of stress that nudges the body to produce more lean mass, improve insulin sensitivity, encourage fat loss, and promote bone building. Stacy Sims says, "It's a different ball game."

Sims is a proponent of swapping out long, slow cardio sessions in favor of shorter bouts of high-intensity work. She says it doesn't mean you can't do your long run or ride, but it shouldn't be the only thing on your training calendar. Endurance training breaks down muscle over time, which, during this phase of life, your body is already prone to do.

Instead, high-intensity interval training (HIIT), where you repeat super-short intervals of work (usually no more than twenty seconds) at near all-out effort followed by a rest period, as well as plyometrics (like box jumps or jump squats) stimulate fast-twitch muscle fibers, the type of muscle fibers that tend to retreat with age. Sims says HIIT also spikes metabolism, which helps clear glucose from the bloodstream and counteract the body's increased insulin resistance. Research indicates that HIIT can reduce total fat, including the visceral kind.

And Sims says don't shy away from strength training. Dumbbells, resistance bands, and bodyweight exercises load tissues and bones to help make up for the lack of estrogen and the hormone's influence on muscles and bone density. Heavy weights force muscles to contract more forcefully, providing a more powerful stimulus to jump-start muscle building and reinforce neuromuscular connections between the brain and muscles. For people who are good candidates, preliminary research suggests that the combination of hormone therapy and strength training could postpone the loss of muscle mass and strength while improving muscle function. Researchers think it's because the estrogen in hormone therapy improves the body's response to training stimuli.

Strength training could potentially reduce the likelihood of injury at a time when injury risk changes. It makes muscles and the surrounding connective tissues more resilient, so they can better support the joints and improve mobility and proprioception. That can mean fewer falls and potential bone injuries, which plague older women. During menopause, not only are bones more fragile, but bone building is impaired, meaning it's harder to repair bones when they do break. Improving muscle mass and function also can help to guard against tendon injuries, like an Achilles tendon rupture. Without the steady supply of estrogen to keep tendons flexible and adaptable, they become stiffer and the risk of injury after menopause increases. We also see this with the rotator cuff tendon in the shoulder; the prevalence of tears is higher in postmenopausal women compared to those who are still menstruating.

When women take these factors into account and work with their changing bodies, they can still perform at a high level no matter their age. Since turning forty-five, ultrarunner Magda Boulet won the Marathon des Sables (a 250-kilometer race run over seven days in Morocco) and the Leadville Trail 100 Run, two prestigious races in the sport. As she's gotten older, she takes her recovery days more seriously (and without guilt) and incorporates more cross-training and strength sessions into her routine. While she may not be as fast as she used to be, she says she still looks "for the gaps and the small opportunities, and they are still out there."

Adventure athlete Rebecca Rusch only started racing mountain bikes when she was thirty-eight years old, often beating competitors who were more than ten years her junior while racking up a slew of titles, including seven-time world champion and four-time Leadville Trail 100 MTB champion. Like many people, Rusch fell off the training wagon for a couple of years in her midforties. It was partly the result of a busy time in her life—she rode the entire length of the Ho Chi Minh Trail in Vietnam, documenting the 1,200-mile journey for the Emmy Award–winning film *Blood Road*, and promoted the project when it was released in 2017. Rusch thinks it also coincided with her perimenopausal period. "I wasn't exercising, and I wasn't activating those hormones," she told me about the period of time after the film's release. "And that was a big wake-up call for me."

Rusch knew that in order to jump-start her training, she needed a new adventure, something she'd never done before. It had to be big and it had to be scary. She needed to feel like a beginner again, she told me. So she decided she would traverse 350 snowy miles of the famous Iditarod Trail in Alaska on her bike as part of the Iditarod Trail

Invitational. It's a race Rusch previously said she'd never do because she's kind of terrified of the extremely cold environment. But that audacious goal prompted her to train consistently, and it paid off. Rusch has since competed in Alaska three times, winning the women's race in 2019 and 2020. In 2021, she finished first in the self-support category. At the age of fifty-two, she said she felt the strongest she's ever felt on an expedition.

Her training volume wasn't massive, like you'd expect when prepping for an ultra-distance race. It did include a lot of power work to prepare her body for the demands of riding a bike loaded down with sixty pounds of gear, which she'd often have to push and lift over the trails. To simulate the friction of riding through the snow, her coach had her ride low-cadence intervals—pedaling slow revolutions of the wheel to build muscular endurance. In the middle, she'd hop off the bike and do a bunch of push-ups and repeat. She incorporated strength training as part of her daily life, similar to Amanda Thebe's concept of "fitness snacking"—doing a few pull-ups while she did the laundry or walking lunges while she walked the dog. "It doesn't have to be like this hour-long session," she says. It's about consistency.

Rusch says her performance on the bike wasn't that different before and after menopause. In 2020, when she tested the amount of power she could sustain on the bike for an hour, her numbers weren't far off from her results in 2013, prior to menopause when she was racing frequently. In some areas, her numbers had actually improved. However, how she felt during and after the sessions was different. "I have to pay attention to what I call the spaces in between," she says, like stretching, hydration, and sleep, which has made a huge difference in how she feels. "It's made me feel good as an athlete, but it's also made me feel good as a human."

Rusch hasn't always talked about age and what it means for an athlete like her. But she was proud of her expeditions in Alaska. "I can do these new amazing things I've never done before and I feel like that has opened a [new] level of confidence in my ability," she says. She wanted to show others that they could be lifelong athletes, that they could discover and open new doors within themselves too.

As the daughters of Title IX—the generation of women who grew up in the wake of the landmark legislation—enter the middle stage of life, they don't want their grandmother's or their mother's menopause experience. They don't want to spend the second half of their lives feeling irrelevant, and they certainly don't want to stop just because their periods do. They want to continue to be active, whether they grew up with sports and fitness as part of the daily fabric of their lives or found their love for movement as an adult. Most importantly, they want information—good information—to understand the transition and their symptoms, as well as ways to manage them and thrive.

With one billion women around the world estimated to be postmenopausal by 2025, menopause is finally having its turn in the spotlight. Books, podcasts, articles, technology, and products are flooding what's considered to be a $600 billion market. There are start-ups like Gennev, whose mission is "to empower women in their post-reproductive years." They not only connect people to healthcare providers in all fifty states who specialize in menopause (many of whom are NAMS-certified) via virtual appointments, but also provide education and community for those stumbling through this experience on their own. "I feel like this generation is pushing it out into the sunshine and being like, okay, we're not going to hide in the corners and whisper about this anymore. We're going to get shit done," Selene Yeager says.

In October 2020, Yeager launched the podcast *Hit Play Not Pause* along with a social media community called Feisty Menopause. Her goal is to normalize conversations about hormones and physiology and show women how they can work with their changing bodies to be fit, active, and strong. The night before the podcast launched, Yeager told me she had a moment of panic. Did she really want to be the face of a menopausal movement? But deep down, she knew how feeling invisible can have a pernicious effect. She herself shrank away from the world around her fiftieth birthday. "If no one is talking about it, everyone is in that dark place by themselves and it becomes a self-fulfilling prophecy," she says.

Once Yeager turned on the mic, active peri- and postmenopausal people emerged from all corners. They shared how they retreated from life and pulled themselves off the start lines of races. They thought they were done pursuing the sports and activities they loved. Through connecting with others going through similar experiences, these women are talking and sharing tips and resources. They're no longer invisible.

While these open conversations, community, and resources are critical, it's only the tip of the iceberg. If we want people to be prepared and supported during and after the menopause transition, we need more research on active and athletic middle-aged populations. A *lot* more research.

It's crucial to investigate active women at different points of the lifespan, including peri- and postmenopause. Sports science researchers now recognize this data chasm, especially as many of the leading scientists in the field find themselves approaching or in the middle of menopause. For instance, how do menopause and the absence of reproductive hormones affect exercise capacity and training adaptation? With the decline in estrogen, does the body start to behave more like a typical male body? What's the influence of MHT on athletic performance? Is it similar to or different from the effect of hormonal contraceptives? And what about alternatives for those who can't tolerate estrogen therapy? While researchers have investigated changes in bone and muscle health during menopause, those studies were conducted primarily with nonactive participants. What are the effects in people who are already active, and can they offset those changes if they have a higher baseline of fitness and strength? Does the decline in estrogen influence risk for injury and conditions like RED-S?

And menopause deserves more than just a moment in the spotlight. We need to reframe the social and cultural narrative around middle age away from the limiting tropes that equate people with their wombs and reproductive capacity. Menopause is a fact of life, neither good nor bad. The cessation of the menstrual cycle doesn't mean you're automatically irrelevant or that you have to suffer alone. "You've got to go through [menopause]. Nobody escapes it. It's just something that's there. And so instead of stressing or being scared about it, be educated about it," Rebecca Rusch says.

Embracing menopause and reclaiming exercise and sports on your own terms requires a bit of a beginner's mindset—approaching your body, fitness, and sports with a sense of curiosity and kindness. That shift in perspective can make the pursuit of an active lifestyle feel invigorating and challenging in a way that your younger self often overlooked and underestimated. It's a new chapter of competition and performance to explore. And like Rusch, you may uncover new reams of potential within yourself too.

CONCLUSION

Beyond the Gap

I n the days leading up to the 2021 NCAA Basketball Tournament, one of the biggest sporting events in the United States, an image circulated on social media. It looked like before-and-after photos from a gym renovation project. The before photo: a single A-frame rack of six pairs of dumbbells and a few black yoga mats set against a gray wall. The after photo: a large, open room divided into multiple weightlifting areas with benches, squat racks, barbells, a variety of plates, and a full range of dumbbells. Except these weren't before-and-after pictures—it was a photo of the weight room inside the women's championship tournament bubble in San Antonio, Texas, and the weight room inside the men's championship tournament bubble in Indianapolis, Indiana.

The photo went viral, and the NCAA had to respond, explaining that the reason for the discrepancy was a matter of space. There just wasn't enough room in the women's bubble to accommodate a bigger weight room. That's when University of Oregon basketball player Sedona Prince posted a video on TikTok that illustrated the absurdity of the NCAA's excuse. She showed the women's practice court, then panned across the space, dipping her camera down to show the "weight room," before revealing an enormous area next to the court that sat empty, save for a few folding chairs. Space, clearly, wasn't the issue.

It wasn't just the blatant disparity that was shocking. It was the fact that the NCAA thought their actions were acceptable, that there was nothing out of the ordinary, that no one would notice. And it wasn't out of the ordinary. The women's tournament also received different and subpar food, swag bags, transportation, and marketing. There wasn't a plan to offer courtesy photos until the Sweet Sixteen, while photos were available for the entirety of the men's tournament, making it immensely more difficult for the media to cover the women's games. The differences in resources between the two tournaments reflected an underlying assumption and attitude that has long pervaded the NCAA: women should be happy they have an opportunity to play and appreciate what they're given—no questions asked.

But by pressing record and harnessing the power of social media, Prince looked directly into the camera and at the NCAA to say that the players weren't going to sit silently while they were disrespected. As Prince says in her video, "If you aren't upset about this problem, then you're a part of it."

Coaches, WNBA players, and others joined the call too. Muffet McGraw, legendary former head coach at Notre Dame women's basketball, tweeted: "We have been fighting this battle for years and frankly, I'm tired of it. Tired of turning on the tv to see 'NCAA basketball tournament' only to realize that of course that means men's. Tired of seeing Twitter accounts called March Madness and Final Four that are run by the NCAA but only cover men's bball. Tired of having to preface everything we do with the word 'Womens' which would be fine if the men had to do the same, but they don't, and when they don't it makes us look like the JV tournament to their event."

The revelations of gender bias at the college basketball tournament, and the accompanying outcry, forced the NCAA's hand. The organization retained the law firm Kaplan, Hecker & Fink LLP to conduct an external review, and the firm released the first of two reports in August 2021, focused on the basketball championship tournaments. (The second report focused on the other eighty-four national championships.) They found that the NCAA spent $35 million more on the men's tournament than the women's event in 2019, the last year for which there were finalized financials at the time. While they tallied up the myriad ways the women's and men's events were different—and there were so many—the firm pointed to something rooted much deeper than just budgets and resources.

In the report, they wrote:

> The primary reason, we believe, is that the gender inequities at the NCAA—and specifically within the NCAA Division I basketball championships—stem from the structure and systems of the NCAA itself, which are designed to maximize the value of and support to the Division I Men's Basketball Championship as the primary source of funding for the NCAA and its membership. The NCAA's broadcast agreements, corporate sponsorship contracts, distribution of revenue, organizational structure, and culture all prioritize Division I men's basketball over everything else in ways that create, normalize, and perpetuate gender inequities. At the same time, the NCAA does not have structures or systems in place to identify, prevent, or address those inequities. The results have been cumulative, not only fostering skepticism and distrust about the sincerity of the NCAA's commitment to gender equity, but also limiting the growth of women's basketball and perpetuating a mistaken narrative that women's basketball is destined to be a "money loser" year after year.

The gender inequity on display at the championship tournament wasn't the fault of the women's tournament committee. It was the result of the larger organization, built to support a specific perspective, one that largely excludes women and people of color and hamstrings any attempts at parity. As Anthony Weems, an assistant professor who studies sports ethics, told *Global Sport Matters,* "These organizational structures were built over time in ways that did not consider diverse viewpoints, that did not consider viewpoints outside of the dominant perspective of mostly white men."

The NCAA's priority during the pandemic was to get the 2021 men's basketball championship up and running, since it's the organization's main moneymaker. On November 16, 2020, they announced that the men's tournament would take place in March 2021, but the women's tournament didn't get the green light until December 14, 2020, a full month after the men. The women's committee was required to plead its case that their tournament should even take place, including a financial review and approvals for additional expenses related to COVID-19— something the men's committee didn't have to do. The extra time and labor required to navigate these additional hoops led to the later approval date and a series of delays that left the women's committee scrambling to plan a major sporting event.

The NCAA story says a lot about collegiate athletics, but it's also a microcosm of what we see in multiple arenas of sports, including the field of sports science. The onus is placed on women to overcome the obstacles inherent in a system that was rigged against them from the get-go. When they don't succeed, they are blamed for their shortcomings. When they do succeed, it tracks as masculine, inappropriate, and a threat. It's an impossible situation.

Like the NCAA, the research ecosystem isn't built to accommodate multiple points of view. It's designed to prioritize and support the understanding of male bodies and men's performance. Biology and physiology become easy scapegoats. Because women's bodies are different, women are vulnerable and should be protected. Because women's bodies are different, women shouldn't play sports. Because women's bodies are different, women are deficient, defective, and therefore not good athletes. Because women's bodies are different, we shouldn't study them.

Since men are positioned as the "perfect" specimen, the default norm by which all else is judged, funding historically flowed more easily to studies investigating male cells, animals, and humans. Since everyone studies male specimens, journal editors and peer reviewers are more familiar with this context and may look upon these studies more favorably for publication. Meanwhile, researchers have had to clear countless hurdles to prove a study involving women is even worthwhile. They may be met with more skepticism when vying for publication because the methodology and research questions are different or less familiar to reviewers. Or, if they're studying a question already investigated in men, the study is deemed redundant. If the studies are published, researchers are expected to preface their work with "women" or "female" in the title. And without a robust infrastructure in place to monitor and promote gender equity in research, we're left with a self-perpetuating cycle where the same assumptions and biases continue to fuel the status quo, unchecked and unacknowledged.

While writing this book, one question kept ringing in the back of my mind: What do we miss when we under-study women? Through the reporting process, the answer became clear. The failure to study girls and women and to recognize their lived experiences has greatly impacted their role, participation, and success in sports, and ultimately their health and well-being.

The system makes it harder for girls and women to get involved and stay involved in physical activity and for athletes to have long, healthy careers. Girls and women are blamed when their athletic progress stalls or they burn out on sports, when they sustain an injury, when they don't return to form quickly after having a child, or when they slow down as they mature. But the truth is that the system hasn't provided girls and women access to the training, protocols, research, and even clothing and gear that might change that trajectory. In its shortsightedness, the system has set unrealistic expectations for women's athleticism and capabilities.

And it's not just a matter of inclusion or diversity. The biases, resource gaps, and structural inequities carry important stakes, ones that affect real human beings. "We end up losing really talented athletes because they have not been looked after properly and they have not been able to train their body to the best of their ability," New Zealand rower Jackie Kiddle told me. "I hate to think that there are athletes out there that can't achieve what they should be able to achieve because of a system that they're in." For other populations marginalized by sex and gender like transgender, nonbinary, and intersex people, the harm runs even deeper and can be detrimental to mental, physical, and emotional health and well-being. Since they don't fit the conventional categories in the binary sports world, their right to play is questioned and debated.

What's incredibly exciting about science is its ability to illuminate the previously unknown. It allows us to examine phenomena we experience in daily life in a structured, methodical way. To understand how the world works a little better. To unlock universal truths.

And what's incredibly frustrating about science is its slow pace, though it's an intrinsic feature of the scientific process. Scientists begin with a research question and slowly construct a base of evidence, one question at a time. They rely on others to validate their findings to ensure hypotheses hold true over the long-term. Not only does it take time to build up sufficient evidence, it takes even more time to translate that science into something practical that a doctor, coach, or physical therapist might use. There's an oft-cited statistic that it takes an average of seventeen years for research findings to be incorporated into clinical practice. That means if a new study on women is released and a girl is born on the same day, she will likely graduate from high school before healthcare and sports professionals start implementing the study's findings. While the step-by-step process ensures the validity of research findings, it also means that science doesn't always keep up with the current times.

Which is where we find ourselves today, and the desire for better information is palpable. Researchers are diligently working to piece together the puzzle on women's hormones, physiology, anatomy, injury, pelvic floor, breasts, pregnancy, the postpartum period, menopause, mental health, and psychology and how all these aspects influence active women and athletes. They've barely scratched the surface.

There's so much work to do to create a sports system that's truly inclusive not only for women but for *all* people. We need more women participants in sports science research as a whole, especially women of color and from non-Western countries, to continue building a robust picture of active and performance-driven women. We need more good-quality studies on how the unique physiology of a woman's body does or doesn't affect exercise capacity, fueling needs, injury risk, endurance capacity, and athletic performance across all stages of a person's life. As we've seen, bodies don't exist in a vacuum. To generate more informed and nuanced recommendations, there should be more interdisciplinary research and collaboration so that we can better understand the full context in which people are active and play sports, from the professional to the recreational level. And researchers need to do a better job of explicitly stating the gender breakdown of participants and explaining how sex-based difference may or may not influence outcome measures—ideally in study abstracts, because not everyone has access to or will read full journal articles.

However, all this work won't move the needle if we don't address the translational research pipeline so that we can get the latest research to where it can make a difference—into the hands of doctors, coaches, physical therapists, athletic training staff, parents, women, and girls. The launch of the Wu Tsai Human Performance Alliance in July 2021 provided a huge shot in the arm for this work. Backed by $220 million from the Joe and Clara Tsai Foundation, the alliance brings together six leading academic institutions in the United States to uncover the principles of peak performance and health, including women's health and sports performance.

What's unique about the alliance is that it isn't solely focused on groundbreaking research. They've made a commitment to translate and test those findings. "There's a back-and-forth where the researchers will talk to the innovation hubs and other partners to understand what are the actual needs, and that will help drive some of the research," Joy Ku, the alliance's education and outreach lead, told me. "But then the outcomes of that research will be translated back and tested" at partnering institutions, including Boston Children's Hospital's Female Athlete Program. They've also committed to bringing the science to larger audiences through educational materials and digital resources. The hope is that by understanding humans at their best, scientists can then apply those principles across the age and athletic spectrum.

It's crucially important to recognize that humans are wonderfully diverse and messy. We aren't just a collection of hormones, bones, muscles, and organs or an amalgamation of data points in a perfectly controlled laboratory setting. While science and research provide important details that can inform decisions about training, fueling, injury prevention, return to sports, and staying active throughout life, finding the best strategy for any one person will require a lot of trial and error. And that requires empowering women and others to understand and reclaim their own bodies.

As our understanding of sex and gender continues to evolve, we need to consider people who are nonbinary, transgender, and have intersex bodies. Current infrastructure and policies seem to question the right of these individuals to exist in a world of sports that is starkly defined by definitive categories of sex and gender. Rules that determine whether they can participate in competitive sports vary from state to state, nation to nation, sport to sport, and across different levels of competition based on physiological and biological criteria like hormone levels. There's minimal research conducted with these populations, yet sports organizations make decisions and recommendations that impact them based on information that may not be appropriate or applicable.

As a result, transgender and intersex people are often barred from playing sports and nonbinary people are forced to pick a sporting category that may not align with their gender identity. WNBA player Layshia Clarendon, who is transgender and nonbinary, told *Sports Illustrated*, "Right now, you either fit in or you get lost. Like, you meet the NCAA or IOC standards that make you eligible to play on the men's or women's side, or you don't. Or you transition and you get lost, forced to move on with your life." If they do participate, they may be subject to harassment, discrimination, and bullying. All these situations can be detrimental to one's physical and mental health.

There's a range of possibilities for what sports can and should look like—we don't have to keep the status quo or blow the whole system up. However, we do need to be willing to have conversations about how to create sports policies that include those who have been left out by design. What's become clear is that addressing the research and resource inequities between women and men and gathering data on a wider variety of people ultimately help us learn more about humans, across the biological sex and gender spectrum.

There's so much good to be found through physical activity and sports, whether it's a fitness class at the YMCA, a high-intensity boot camp at a boutique studio, a slow flow yoga class, a pickup basketball game, or an Olympic final. In my interviews, I often asked people about their relationship to sports and why it's important to them. Winning is great, but over and over, people said they love playing sports and being physically active—moving their bodies brings them joy. It's taught them confidence and self-worth, given them freedom, and expanded what they believe is possible. It's formed meaningful lifelong friendships and mentorships.

As I was revising this book in January 2022, I was thinking about how far women have come in sports and how much farther we can go. As if on cue, my Twitter feed lit up. On January 16 in Houston, Texas, thirty-eight-year-old Sara Hall finished second in the half-marathon in 1:07:15, setting a new American record in the distance. About an hour later, thirty-seven-year-old Keira D'Amato crossed the finish line of the marathon in 2:19:12, breaking the tape and the American marathon record set by Deena Kastor, which had stood since 2006.

While it was a great day for running overall, it was a powerful one for women. Both athletes are parents and well past what's traditionally considered one's athletic prime. Hall is gritty, someone who has persevered in the sport for more than fifteen years. While she's notched world-class performances in multiple distances, she's had her fair share of disappointments too. She never made an Olympic team, has come up short in some big races, and has suffered injuries. D'Amato's story is the more unlikely one. She quit competitive running after college due to injury, got married, had two kids, and worked full-time as a Realtor. She only returned to the sport in 2016, after a seven-year hiatus, to help cope with the stress of life. In the past few years, she's emerged as one of the best distance runners, and she signed her first pro contract in February 2021.

As cliché as it sounds, both women prove that age really is just a number. In an interview with the *Today* show, D'Amato said part of what drives her is her curiosity to find out what she is capable of, and she doesn't believe she's found her full potential yet. She also tweeted, "After setting the American Record in the Marathon, the thing I am most excited about is some girl/woman saw it and thought, 'I can do that.' I know they will . . . and I'll be rooting for them."

As January came to a close, women demonstrated again that, when given an opportunity, they'll rise to the occasion and shine. On the morning of January 30, 2022, pro surfers Sally Fitzgibbons, Courtney Conlogue, and Molly Picklum paddled into the water off the North Shore of Oahu in Hawaii for the first heat of the Billabong Pro Pipeline. It was a historic moment—the first time that a full Women's Championship Tour event was held at the storied surf break, known for its pristine hollow barrels and waves so heavy that the beach shakes. (Men have competed at Pipe since 1971.)

But the most exciting part were the new faces dotting the lineup—five rookies, four of whom were teenagers. This new generation of surfers is dynamic and have put world champions like Carissa Moore and Stephanie Gilmore on notice. Doing aerials, pulling into huge barrels, and blowing their fins out of the water—things we're used to seeing on the men's side of the tour—aren't just moves they aspire to do. They are maneuvers these athletes are doing right now. Watching these incredibly talented young women was exciting, and it looked like they were having so much fun.

At both ends of the age spectrum, women are shaking off the expectations of the past and pushing the boundaries of what can be. And when left unconstrained to explore their potential, they're blazing a new path and model of athleticism and performance, one that's valid in its own right and doesn't constantly refer back to what men have accomplished.

Because what are the limits? When we consider the full complexity of human experience beyond just cisgender men, when we give equal weight to those experiences and acknowledge them, when we provide equal access to competition opportunities, resources, and scientific knowledge, when we have a system of sports and science that is truly inclusive, we just might find that all people—at any age and level—can be their best in the pursuit of physical activity and sports. When we have a more inclusive frame of reference for what it means to be an active person or an athlete, we may discover that there are so many more possibilities than we previously thought.

And we might also find that there is no limit.

Acknowledgments

While I spent countless hours staring at a blinking cursor, wondering if the words would ever materialize, I had an incredible support crew behind me. This book came into being with the encouragement and kindness of so many people.

First and foremost, I am grateful for my editor, Courtney Young, who has been a tremendous partner in this project. After our first phone call, I knew I wanted to work with Courtney. I knew she wouldn't take it easy on me, and she didn't. She pushed me, asked all the hard questions, and didn't let me take the simple way out. Her thoughtful guidance, feedback, and edits helped turn a rambling first draft into a book. Thank you for believing in this book, and me, from the beginning. Thank you, Jacqueline Shost, for your insightful comments and for keeping everything on schedule. To the rest of the team at Riverhead—Katie Hurley, Amy Ryan, Will Jeffries, Carla Benton, Rebecca Reisert, Shailyn Tavella, Kitanna Hiromasa, Ashley Sutton, and Katie Vaughn—thank you for making this process seamless. I feel so fortunate to have found my publishing home with you. Huge thanks to Pete Garceau and Helen Yentus for the beautiful cover design and to Amanda Dewey for the interior design.

This book would not exist without a few key people, especially my agent Susan Canavan, who championed this idea from the very beginning and convinced me that yes, I really did want to write a book. Immense thanks to Molly Mirhashem, my editor at *Outside*. She gave me the opportunity to explore a wide range of issues related to women athletes, sports science, and performance, which sparked many of the questions that led to this book. She helped me find my voice and encouraged me to use it confidently, ultimately making me a better writer. Special thanks to Kaelyn Lynch for meticulously fact-checking every page of the manuscript and to Maura Fox and Claire Hyman for sorting through piles of scientific papers and other research. Your help was invaluable.

As someone who has a tendency to over-research and who likes to show her work, it pains me that I could not cite or quote all of the people I interviewed because each of those conversations pushed my thinking further. Thank you to everyone who gave their time to speak with me, especially Kate Ackerman, Tony Hackney (who sometimes felt like a guardian angel on my shoulder, sending words of encouragement when I most needed them), Emily Kraus, Kirsty Elliott-Sale, Kelly McNulty, Stacy Sims, Trent Stellingwerff, and Abby Bales. Thank you to the pioneers in the field, like Carole Oglesby, Jan Todd, and Aurelia Nattiv, for sharing your stories with me.

I am forever indebted to Michelle Hamilton and Theodora Blanchfield, who have both been lifelines to me during the book writing process. Thank you to Lauren Fleshman and Marianne Elliott for teaching me to listen to my "inner sweetheart" and to my Wilder sisters for always holding space for me: Petra Moll, Sarah Meyer Tapia, Elizabeth Ewens, Kim Barman, Kristjana Cook, Erin Good, Sarah Overpeck, Lauren Udwari, Karen Principato, Hallie Caplan, and Ali Glenesk. Special thanks to Erin Lentz, Sarah Canney, Steph Creaturo, Amanda Loudin, Kim McCreight, Elizabeth Carey, Cindy Kuzma, Laura Tillman, Ali Feller, and Tina Muir for your tremendous support.

My deepest thanks and gratitude to my family for always being in my corner: Mom, Mitrofan and Jadwiga, Beck and James, Ron and Jane, Olga and Chris, Tracy Wang, Alison Wang, and all the kids. Yes, I'm finally done.

To Jasper and Everett: For your patience, love, and understanding when I was holed up in my office. Thank you for keeping me on track with your checklists, surrounding me with the best coworkers, and giving the best hugs and advice.

To Ed: For your unwavering patience and support. Thank you for being the best first reader and for making sure it wasn't too inside baseball. Thank you for lifting me up when my confidence faltered, constantly reminding me that I'm more than capable of writing this book. I couldn't have done this without you.

Notes

During the course of reporting this book, I read several hundred scientific papers and conducted more than 140 interviews. The citations that appear here represent some of the key articles, papers, and studies that informed my thinking. The majority of quotes in the book are from interviews I conducted; quotes from other sources are identified here.

Introduction: Mind the Gap

xiii **reporting on an article:** Christine Yu, "Where Are the Women in Sports Science Research?," *Outside*, November 8, 2018, https://www.outsideonline.com/health/training-performance/where-are-women-sports-science-research/.

xiv **number of girls and women:** "2018–2019 High School Athletics Participation Survey," National Federation of State High School Associations, https://www.nfhs.org/media/1020412/2018-19_participation_survey.pdf; Nancy Armour et al., "'They've Had 50 Years to Figure It Out': Title IX Disparities in Major College Sports Haven't Gone Away," *USA Today*, March 30, 2022, https://www.usatoday.com/in-depth/news/investigations/2022/03/30/title-ix-50th-anniversary-women-short-changed-major-college-sports/7090806001/.

xiv **fields of exercise and sports science:** Emma S. Cowley et al., "'Invisible Sportswomen': The Sex Data Gap in Sport and Exercise Science Research," *Women in Sport and Physical Activity Journal* 29, no. 2 (2021): 146–51, https://doi.org/10.1123/wspaj.2021-0028.

xvi **Pro runner Leah Falland:** Leah Falland, "Hey, Struggling Runner—," *Medium*, November 6, 2018, https://medium.com/@LeahKayO/hey-struggling-runner-7ab9d22155da.

xviii **the field of sports science began to change:** Key sources include interviews with Jan Todd and Carole Oglesby. See also Jason P. Shurley, Jan Todd, and Terry Todd, *Strength Coaching in America: A History of the Innovation That Transformed Sports* (Austin: University of Texas Press, 2019), 79, 88–89, 134–35; Jaime Schultz, "A History of Kinesiology," in *Foundations of Kinesiology*, ed. Carole A. Oglesby, Kim Henige, Douglas W. McLaughlin, and Belinda Stillwell (Burlington, MA: Jones and Bartlett Learning, 2022), 44–48.

xxi **the 1900 Olympics in Paris:** "Gender Equality Through Time: At the Olympic Games," International Olympic Committee, https://olympics.com/ioc/gender-equality/gender-equality-through-time/at-the-olympic-games.

xxii **a video op-ed:** Mary Cain, "I Was the Fastest Girl in America, Until I Joined Nike," *The New York Times*, November 7, 2019, https://www.nytimes.com/2019/11/07/opinion/nike-running-mary-cain.html.

1. Where Are All the Women?

1 **mountain biker Kate Courtney:** Details about Kate Courtney are drawn from an interview with Courtney and sources including Neal Rogers, "American Muscle," *The Red Bulletin*, July 23, 2020, https://www.redbull.com/us-en/theredbulletin/kate-courtney-and-chloe-dygert-american-muscle; Matt Ruby and Noah Throop, "Kate Courtney Is Creating Pathways for American Cyclists," *The New York Times*, July 25, 2021, https://www.nytimes.com/interactive/2021/07/26/sports/olympics/kate-courtney-mountain-bike.html.

3 **Courtney shared her concerns:** Kate Courtney (@kateplusfate), "If you can see it, you can be it . . . ," Instagram photo, January 25, 2022, https://www.instagram.com/p/CZKc5E2Plf3/.

3 **Modern sports developed:** Jason P. Shurley, Jan Todd, and Terry Todd, *Strength Coaching in America: A History of the Innovation That Transformed Sports* (Austin: University of Texas Press, 2019), 22; Michael A. Messner, *Power at Play: Sports and the Problem of Masculinity* (Boston: Beacon Press, 1992), 10–19; Ryan Swanson, "Teddy Roosevelt and the Strenuous Life," *American Heritage* 64, no. 1 (Winter 2020), https://www.americanheritage.com/teddy-roosevelt-and-strenuous-life.

4 **American youth were becoming less physically fit:** Robert H. Boyle, "The Report That Shocked the President," *Sports Illustrated*, August 15, 1955, https://vault.si.com/vault/1955/08/15/the-report-that-shocked-the-president; John F. Kennedy, "The Soft American," *Sports Illustrated*, December 26, 1960, https://www.jfklibrary.org/asset-viewer/archives/JFKPOF/094/JFKPOF-094-003.

4 **the Harvard Fatigue Laboratory:** Shurley, Todd, and Todd, *Strength Coaching in America*, 34; Andi Johnson, "'They Sweat for Science': The Harvard Fatigue Laboratory and Self-Experimentation in American Exercise Physiology," *Journal of the History of Biology* 48, no. 3 (2015): 425–54, https://www.jstor.org/stable/43863409; Robin Wolfe Scheffler, "The Power of Exercise and the Exercise of Power: The Harvard Fatigue Laboratory, Distance Running, and the Disappearance of Work, 1919–1947," *Journal of the History of Biology* 48, no. 3 (August 1, 2015): 391–423, https://doi.org/10.1007/s10739-014-9392-1.

5 **At the collegiate level:** "NCAA Sports Sponsorship and Participation Rates Database," NCAA, https://www.ncaa.org/sports/2018/10/10/ncaa-sports-sponsorship-and-participation-rates-database.aspx.

6 **conference on women in sports:** Louise Burke, "Why We Need to Undertake More Research on Female Athletes," presentation, Women in Sport and Exercise Academic Network (WiSEAN), virtual, September 8, 2020.

6 **USA Today analyzed:** Nancy Armour et al., "'They've Had 50 Years to Figure It Out': Title IX Disparities in Major College Sports Haven't Gone Away," *USA Today*, March 30, 2022, https://www.usatoday.com/in-depth/news/investigations/2022/03/30/title-ix-50th-anniversary-women-short-changed-major-college-sports/7090806001/.

7 **global sports market:** Paul Lee, Kevin Westcott, Izzy Wray, and Suhas Raviprakash, "Women's Sports Gets Down to Business: On Track for Rising Monetization," *Deloitte Insights*, December 7, 2020, https://www2.deloitte.com/xe/en/insights/industry/technology/technology-media-and-telecom-predictions/2021/womens-sports-revenue.html.

7 **the creation of the "standard man":** G. E. McMurtrie, ed., *"Report of the United Kingdom Delegation,"* Permissible Doses Conference, Chalk River, Ontario, Canada, September 29–30, 1949, May 1950, https://nsarchive2.gwu.edu/radiation/dir/mstreet/commeet/meet6/brief6/tab_n/br6n2a.txt; International Commission on Radiological Protection, *Report on the Task Group on Reference Man*, ICRP Publication 23 (Oxford: Pergamon Press, 1975), 1–5.

8 **case study of sports-related courses:** Philippa Velija and Catherine Phipps, "Gendered Curricula and Female Students' Experiences on University Sport Courses," presentation, Women in Sport and Exercise Academic Network (WiSEAN), virtual, September 8, 2020.

9 **biggest consideration is the menstrual cycle:** Stacey Emmonds, Omar Heyward, and Ben Jones, "The Challenge of Applying and Undertaking Research in Female Sport," *Sports Medicine—Open* 5, no. 1 (December 12, 2019): 51, https://doi.org/10.1186/s40798-019-0224-x.

10 **Body temperature changes:** Charles L. Buxton and William B. Atkinson, "Hormonal Factors Involved in the Regulation of Basal Body Temperature During the Menstrual Cycle

and Pregnancy," *The Journal of Clinical Endocrinology and Metabolism* 8, no. 7 (July 1, 1948): 544–49, https://doi.org/10.1210/jcem-8-7-544.

11 **Volunteer bias:** James Nuzzo, "Volunteer Bias and Female Participation in Exercise and Sports Science Research," *Quest* 73, no. 1 (2021): 82–101, https://doi.org/10.1080/00336297.2021.1875248.

12 **When Ken O'Halloran reviewed:** Ken D. O'Halloran, "Mind the Gap: Widening the Demographic to Establish New Norms in Human Physiology," *The Journal of Physiology* 598, no. 15 (2020): 3045–47, https://doi.org/10.1113/JP279986.

12 **data gap becomes even more skewed:** Joseph T. Costello, Francois Bieuzen, and Chris M. Bleakley, "Where Are All the Female Participants in Sports and Exercise Medicine Research?," *European Journal of Sport Science* 14, no. 8 (2014): 847–51, https://doi.org/10.1080/17461391.2014.911354; Emma S. Cowley et al., "'Invisible Sportswomen': The Sex Data Gap in Sport and Exercise Science Research," *Women in Sport and Physical Activity Journal* 29, no. 2 (2021): 146–51, https://doi.org/10.1123/wspaj.2021-0028.

13 **publishing in the sports science space:** Elena Martínez-Rosales et al., "Representation of Women in Sport Sciences Research, Publications, and Editorial Leadership Positions: Are We Moving Forward?," *Journal of Science and Medicine in Sport* 24, no. 11 (November 1, 2021): 1093–97, https://doi.org/10.1016/j.jsams.2021.04.010; Aamir Raoof Memon et al., "Where Are Female Editors from Low-Income and Middle-Income Countries? A Comprehensive Assessment of Gender, Geographical Distribution and Country's Income Group of Editorial Boards of Top-Ranked Rehabilitation and Sports Science Journals," *British Journal of Sports Medicine* 56, no. 8 (2022): 458–68, https://doi.org/10.1136/bjsports-2021-105042.

14 **Among the trailblazers:** "Dr. Barbara Drinkwater, Ph.D., FACSM—ACSM President 1988–1989," interview, ACSM's Distinguished Leaders in Sports Medicine and Exercise Science, March 19, 2014, video, https://youtu.be/qZc0dWw3qLI.

16 **Drinkwater published two seminal studies:** Barbara L. Drinkwater et al., "Bone Mineral Content of Amenorrheic and Eumenorrheic Athletes," *The New England Journal of Medicine* 311, no. 5 (1984): 277–81, https://doi.org/10.1056/NEJM198408023110501; Barbara L. Drinkwater et al., "Bone Mineral Density After Resumption of Menses in Amenorrheic Athletes," *JAMA* 256, no. 3 (July 18, 1986): 380–82, https://doi.org/10.1001/jama.1986.03380030082032.

17 **first paper on the triad:** Kimberly K. Yeager et al., "The Female Athlete Triad: Disordered Eating, Amenorrhea, Osteoporosis," *Medicine and Science in Sports and Exercise* 25, no. 7 (July 1993): 775–77, https://doi.org/10.1249/00005768-199307000-00003.

18 **shifting infrastructure and public policies:** Katherine A. Liu and Natalie A. Dipietro Mager, "Women's Involvement in Clinical Trials: Historical Perspective and Future Implications," *Pharmacy Practice* 14, no. 1 (2016), https://doi.org/10.18549/PharmPract.2016.01.708.

20 **example of beetroot juice:** Kate A. Wickham and Lawrence L. Spriet, "No Longer Beeting Around the Bush: A Review of Potential Sex Differences with Dietary Nitrate Supplementation," *Applied Physiology, Nutrition, and Metabolism* 44, no. 9 (2019), https://doi.org/10.1139/apnm-2019-0063; Vikas Kapil et al., "Sex Differences in the Nitrate-Nitrite-NO• Pathway: Role of Oral Nitrate-Reducing Bacteria," *Free Radical Biology and Medicine* 126 (October 2018): 113–21, https://doi.org/10.1016/j.freeradbiomed.2018.07.010.

21 **propose common language:** Kirsty J. Elliott-Sale et al., "Methodological Considerations for Studies in Sport and Exercise Science with Women as Participants: A Working Guide for Standards of Practice for Research on Women," *Sports Medicine* 51 (2021): 843–61, https://doi.org/10.1007/s40279-021-01435-8.

2. More Than Just Hormonal

24 **Greek philosopher Plato:** Plato, *The Republic*, trans. Benjamin Jowett (Oxford: The Clarendon Press, 1888), 143.

24 **Roman satirist Juvenal:** *Juvenal and Persius*, trans. G. G. Ramsay (New York: G. P. Putnam's Sons, 1918), 103.

24 **the Heraea or Heraean Games:** Key sources for the history of women and physical

activity include an interview with Jan Todd; Jan Todd, "'As Men Do Walk a Mile, Women Should Talk an Hour . . . 'Tis Their Exercise' and Other Pre-Enlightenment Thought on Women and Purposive Training," *Iron Game History* 7, nos. 2–3 (July 2002): 56–70, https://starkcenter.org/igh_article/igh0703j/.

25 **In the nineteenth century:** Patricia Vertinsky, "Exercise, Physical Capability, and the Eternally Wounded Woman in Late Nineteenth Century North America," *Journal of Sport History* 14, no. 1 (Spring 1987): 7, 9–11, https://www.jstor.org/stable/43609324.

25 **In his 1873 book:** Edward H. Clarke, *Sex in Education; or, A Fair Chance for Girls* (Boston: James R. Osgood and Company, 1873): 125–32.

26 **Senda Berenson Abbott:** H. Grace Shymanski, "Battling for the Hardwood: The Early History of Women's Basketball at Indiana University, 1890–1928," *Indiana Magazine of History* 114, no. 1 (March 2018): 44, https://doi.org/10.2979/indimagahist.114.1.02.

26 **a modern Olympics:** Gertrud Pfister, "Women and the Olympic Games," in *Women in Sport,* ed. Barbara Drinkwater (Oxford: Blackwell Science Ltd., 2000), 3–19; Martin Polley, "Sport, Gender and Sexuality at the 1908 London Olympic Games," in *Routledge Handbook of Sport, Gender and Sexuality,* ed. Jennifer Hargreaves and Eric Anderson (Abingdon: Routledge, February 2014): 30–31.

27 **had a different perspective:** William Shirer, "5 Women Track Stars Collapse in Olympic Race," *Chicago Tribune,* August 3, 1928; "In Amsterdam in 1928, Lina Radke Was the First Female Olympic 800m Champion, but . . . ," International Olympic Committee, https://olympics.com/en/news/in-amsterdam-in-1928-lina-radke-was-the-first-female-olympic-800m-champion-but.

28 **swimmer Greta Andersen:** "The Day Greta Andersen Needed to Be Rescued from Drowning," International Olympic Committee, https://olympics.com/en/news/the-day-that-greta-andersen-needed-to-be-rescued-x2209; Brent Rutemiller, "Female Olympic Champions Talk Candidly About Menstrual Cycles During Competition," *Swimming World,* March 3, 2015, https://www.swimmingworldmagazine.com/news/female-olympic-champions-talk-candidly-about-menstrual-cycles-during-competitio/.

29 **barred from ski jumping:** Brian Mann, "Women Lobby for Olympic Ski Jumping Event," NPR, November 14, 2005, https://www.npr.org/templates/story/story.php?storyId=5011904.

29 **equal pay lawsuit in 2020:** Kevin Draper and Andrew Das, "'Blatant Misogyny': U.S. Women Protest, and U.S. Soccer President Resigns," *The New York Times,* March 12, 2020, https://www.nytimes.com/2020/03/12/sports/soccer/uswnt-equal-pay.html.

29 **women play a best-of-three-sets match format:** Lindsay Gibbs, "Why Women Don't Play Best-of-Five Matches at Grand Slams," *Think Progress,* May 27, 2016, https://archive.thinkprogress.org/why-women-dont-play-best-of-five-matches-at-grand-slams-6458f5b803df/.

30 ***The Question of Rest:*** Mary Putnam Jacobi, *The Question of Rest for Women During Menstruation* (New York: G. P. Putnam's Sons, 1877), 3, 17, 27, 115–67, 227, http://resource.nlm.nih.gov/67041010R.

31 **In Germany, another group:** Gertrud Pfister, "The Medical Discourse on Female Physical Culture in Germany in the 19th and Early 20th Centuries," *Journal of Sport History* 17, no. 2 (1990): 183–98.

32 **Barbara Drinkwater began her research:** "Dr. Barbara Drinkwater, Ph.D., FACSM—ACSM President 1988–1989," interview, ACSM's Distinguished Leaders in Sports Medicine and Exercise Science, March 19, 2014, video, https://youtu.be/qZc0dWw3qLI.

33 **In a 1977 paper:** B. L. Drinkwater et al., "Heat Tolerance of Female Distance Runners," *Annals of the New York Academy of Sciences* 301, no. 1 (1977): 777–92, https://doi.org/10.1111/j.1749-6632.1977.tb38246.x.

34 **distinct physiological and biological characteristics:** Paul Ansdell et al., "Physiological Sex Differences Affect the Integrative Response to Exercise: Acute and Chronic Implications," *Experimental Physiology* 105, no. 12 (2020): 2007–21, https://doi.org/10.1113/EP088548.

35 **important biological rhythms:** Hope C. Davis and Anthony C. Hackney, "The Hypothalamic-Pituitary-Ovarian Axis and Oral Contraceptives: Regulation and Function," in *Sex Hormones, Exercise and Women,* ed. Anthony C. Hackney (Cham, Switzerland: Springer International Publishing, 2017), 1–17.

36 **Estrogen levels can increase fivefold:** Kirsty J. Elliott-Sale et al., "Methodological Considerations for Studies in Sport and Exercise Science with Women as Participants: A

Working Guide for Standards of Practice for Research on Women," *Sports Medicine* 51 (2021): 843–61, https://doi.org/10.1007/s40279-021-01435-8.

36 **The reality is messier:** Jonathan R. Bull et al., "Real-World Menstrual Cycle Characteristics of More than 600,000 Menstrual Cycles," *NPJ Digital Medicine* 2, no. 1 (August 27, 2019): 1–8, https://doi.org/10.1038/s41746-019-0152-7.

37 **these other jobs can influence:** Naama W. Constantini, Gal Dubnov, and Constance M. Lebrun, "The Menstrual Cycle and Sport Performance," *Clinics in Sports Medicine* 24, no. 2 (April 1, 2005): e51–82, https://www.sportsmed.theclinics.com/article/S0278-5919(05)00004 -9/fulltext.

37 **breast support is a big deal:** Nicola Brown, Jenny Burbage, and Joanna Wakefield-Scurr, "Sports Bra Use, Preferences and Fit Issues Among Exercising Females in the US, the UK and China," *Journal of Fashion Marketing and Management* 25, no. 3 (2021): 511–27, https:// doi.org/10.1108/JFMM-05-2020-0084; Brooke R. Brisbine et al., "Breast Pain Affects the Performance of Elite Female Athletes," *Journal of Sports Sciences* 38, no. 5 (2020): 528–33, https://doi.org/10.1080/02640414.2020.1712016.

38 **pelvic floor dysfunction can affect:** Elena Sonsoles Rodríguez-López et al., "Prevalence of Urinary Incontinence Among Elite Athletes of Both Sexes," *Journal of Science and Medicine in Sport* 24, no. 4 (2021): 338–44, https://doi.org/10.1016/j.jsams.2020.09.017.

38 **leaky bladder, prolapse, and pelvic pain:** Jodie G. Dakic et al., "Effect of Pelvic Floor Symptoms on Women's Participation in Exercise: A Mixed-Methods Systematic Review with Meta-Analysis," *Journal of Orthopaedic and Sports Physical Therapy* 51, no. 7 (2021): 345–61, https://doi.org/10.2519/jospt.2021.10200.

39 **so-called sex chromosomes:** Manoush Zomorodi, "Molly Webster: Is Our Definition of 'Sex Chromosomes' Too Narrow?," May 8, 2020, in *TED Radio Hour*, podcast, https://www .npr.org/transcripts/852217439.

40 **People with intersex bodies:** Amanda Montañez, "Beyond XX and XY: The Extraordinary Complexity of Sex Determination," *Scientific American*, September 1, 2017, https://www .scientificamerican.com/article/beyond-xx-and-xy-the-extraordinary-complexity-of-sex -determination/.

3. Period Power

43 **considered a vital sign:** "Menstruation in Girls and Adolescents: Using the Menstrual Cycle as a Vital Sign," Committee Opinion no. 651, *Obstetrics and Gynecology* 126, no. 6 (December 2015): e143–46, https://doi.org/10.1097/AOG.0000000000001215.

44 **The menstrual cycle is regulated:** Hope C. Davis and Anthony C. Hackney, "The Hypothalamic-Pituitary-Ovarian Axis and Oral Contraceptives: Regulation and Function," in *Sex Hormones, Exercise and Women*, ed. Anthony C. Hackney (Cham, Switzerland: Springer International Publishing, 2017), 1–17.

44 **So can stress:** S. N. Kalantaridou et al., "Stress and the Female Reproductive System," *Journal of Reproductive Immunology* 62, no. 1 (2004): 61–68, https://doi.org/10.1016/j.jri .2003.09.004.

44 **for active people:** M. J. De Souza et al., "High Prevalence of Subtle and Severe Menstrual Disturbances in Exercising Women: Confirmation Using Daily Hormone Measures," *Human Reproduction* 25, no. 2 (2010): 491–503, https://doi.org/10.1093/humrep/dep411.

45 **a cascade of problems:** M. L. Rencken, C. H. Chesnut, and B. L. Drinkwater, "Bone Density at Multiple Skeletal Sites in Amenorrheic Athletes," *JAMA* 276, no. 3 (1996): 238–40, https://doi.org10.1001/jama.1996.03540030072035; Kathryn E. Ackerman et al., "Bone Microarchitecture Is Impaired in Adolescent Amenorrheic Athletes Compared with Eumenorrheic Athletes and Nonathletic Controls," *The Journal of Clinical Endocrinology and Metabolism* 96, no. 10 (October 2011): 3123–33, https://doi.org/10.1210/jc.2011-1614; Anette Rickenlund et al., "Amenorrhea in Female Athletes Is Associated with Endothelial Dysfunction and Unfavorable Lipid Profile," *The Journal of Clinical Endocrinology and Metabolism* 90, no. 3 (March 1, 2005): 1354–59, https://doi.org/10.1210/jc.2004-1286; Åsa B. Tornberg et al., "Reduced Neuromuscular Performance in Amenorrheic Elite Endurance Athletes," *Medicine and Science in Sports and Exercise* 49, no. 12 (December 2017): 2478–85, https://doi.org/10.1249/MSS.00000000 00001383.

45 **thirteen young elite runners:** Johanna K. Ihalainen et al., "Body Composition, Energy Availability, Training, and Menstrual Status in Female Runners," *International Journal of Sports Physiology and Performance* 16, no. 7 (March 9, 2021): 1043–48, https://doi.org/10.1123 /ijspp.2020-0276.

46 **Oral contraceptive pills:** Davis and Hackney, "The Hypothalamic-Pituitary-Ovarian Axis," 1–17.

46 **choose to use hormonal contraceptives:** Daniel Martin et al., "Period Prevalence and Perceived Side Effects of Hormonal Contraceptive Use and the Menstrual Cycle in Elite Athletes," *International Journal of Sports Physiology and Performance* 13, no. 7 (2018): 926–32, https://doi.org/10.1123/ijspp.2017-0330.

47 **research on the effect of hormonal contraceptives:** Brianna Larsen et al., "Inflammation and Oral Contraceptive Use in Female Athletes Before the Rio Olympic Games," *Frontiers in Physiology* 11 (2020), https://doi.org/10.3389/fphys.2020.00497; Clare Minahan et al., "Response of Women Using Oral Contraception to Exercise in the Heat," *European Journal of Applied Physiology* 117, no. 7 (July 1, 2017): 1383–91, https://doi.org/10.1007/s00421 -017-3628-7.

47 **how they feel while active:** Georgie Bruinvels et al., "Prevalence and Frequency of Menstrual Cycle Symptoms Are Associated with Availability to Train and Compete: A Study of 6812 Exercising Women Recruited Using the Strava Exercise App," *British Journal of Sports Medicine* 55, no. 8 (2021): 438–43, https://doi.org/10.1136/bjsports-2020-102792.

48 **Danielle Collins experienced years:** Courtney Nguyen, "Champions Corner: Collins Unleashes the Best Tennis of Her Career After Life-Changing Surgery," *WTA Tour* (blog), August 9, 2021, https://www.wtatennis.com/news/2211449/champions-corner-collins-unleashes-the -best-tennis-of-her-career-after-life-changing-surgery.

48 **elite athletes noted side effects:** Guro S. Solli et al., "Changes in Self-Reported Physical Fitness, Performance, and Side Effects Across the Phases of the Menstrual Cycle Among Competitive Endurance Athletes," *International Journal of Sports Physiology and Performance* 15, no. 9 (September 21, 2020): 1324–33, https://doi.org/10.1123/ijspp.2019-0616.

49 **X Games gold medalist Mariah Duran:** FitrWoman (@fitrwoman), "Live Q&A with Mariah Duran . . . ," Instagram video, October 22, 2020, https://www.instagram.com/p /CGp4g2iniHG/.

50 **first described in a scientific study:** Robert T. Frank, "The Hormonal Causes of Premenstrual Tension," *Archives of Neurology and Psychiatry* 26, no. 5 (1931): 1053–57, https:// doi.org/10.1001/archneurpsyc.1931.02230110151009.

50 **Fu Yuanhui squatted down:** Tom Phillips, "'It's Because I Had My Period': Swimmer Fu Yuanhui Praised for Breaking Taboo," *The Guardian*, August 15, 2016, https://www.the guardian.com/sport/2016/aug/16/chinese-swimmer-fu-yuanhui-praised-for-breaking -periods-taboo.

51 **Japanese swimmer Hanae Ito:** Mai Yoshikawa, "Retired Olympian Reshaping Conversation Around Athletes' Periods," *Kyodo News*, November 11, 2020, https://english.kyodonews .net/news/2020/11/8b2478d4b9d4-feature-retired-olympian-reshaping-conversation -around-athletes-periods.html.

51 **Lonah Chemtai Salpeter took to Facebook:** Lonah-Chemtai Salpeter, "I cannot hide that I love challenges . . . ," Facebook, August 7, 2021, https://www.facebook.com/LCSAL PETER/posts/316169079984689.

51 **Charlton Athletic Women's Football Club:** Guy Pitchers, "The Effects of Menstrual and Contraceptive Cycle Monitoring in Elite Soccer," presentation, Women in Sport and Exercise Academic Network (WiSEAN), virtual, September 7, 2020.

52 **an estimated fifty million people:** Donna Rosato, "What Your Period Tracker App Knows About You," *Consumer Reports*, January 28, 2020, https://www.consumerreports.org /health-privacy/what-your-period-tracker-app-knows-about-you-a8701683935/.

53 **estrogen may boost muscle gain:** Belinda Thompson et al., "The Effect of the Menstrual Cycle and Oral Contraceptives on Acute Responses and Chronic Adaptations to Resistance Training: A Systematic Review of the Literature," *Sports Medicine* 50, no. 1 (January 2020): 171–85, https://doi.org/10.1007/s40279-019-01219-1; Eunsook Sung et al., "Effects of Follicular Versus Luteal Phase-Based Strength Training in Young Women," *SpringerPlus* 3, 668 (November 11, 2014), https://doi.org/10.1186/2193-1801-3-668.

53 **exercise might seem different:** Michelle C. Venables, Juul Achten, and Asker E. Jeukendrup, "Determinants of Fat Oxidation During Exercise in Healthy Men and Women: A Cross-Sectional Study," *Journal of Applied Physiology* 98, no. 1 (January 1, 2005): 160–67, https://doi.org/10.1152/japplphysiol.00662.2003; Stacy T. Sims, Laura Ware, and Emily R. Capodilupo, "Patterns of Endogenous and Exogenous Ovarian Hormone Modulation on Recovery Metrics Across the Menstrual Cycle," *BMJ Open Sport and Exercise Medicine* 7, no. 3 (2021), https://doi.org/10.1136/bmjsem-2021-001047.

55 **researchers in the United Kingdom:** Kelly Lee McNulty et al., "The Effects of Menstrual Cycle Phase on Exercise Performance in Eumenorrheic Women: A Systematic Review and Meta-Analysis," *Sports Medicine* 50 (2020): 1813–27, https://doi.org/10.1007/s40279-020-01319-3.

55 **oral contraceptive pill on exercise performance:** Kirsty J. Elliott-Sale et al., "The Effects of Oral Contraceptives on Exercise Performance in Women: A Systematic Review and Meta-Analysis," *Sports Medicine* 50 (2020): 1785–1812, https://doi.org/10.1007/s40279-020-01317-5.

56 **lack of agreed-upon standards:** Kirsty J. Elliott-Sale et al., "Methodological Considerations for Studies in Sport and Exercise Science with Women as Participants: A Working Guide for Standards of Practice for Research on Women," *Sports Medicine* 51 (2021): 843–61, https://doi.org/10.1007/s40279-021-01435-8.

58 **judging from the headlines:** For example, Kieran Pender, "Ending Period 'Taboo' Gave USA Marginal Gain at World Cup," *The Telegraph*, July 13, 2019, https://www.telegraph.co.uk/world-cup/2019/07/13/revealed-next-frontier-sports-science-usas-secret-weapon-womens/.

59 **the brainchild of Dawn Scott:** The primary source on the USWNT's high-performance strategy was an interview with Dawn Scott.

4. Fast Fuel

64 **their top elite women rowers:** Interview with Jackie Kiddle; Suzanne McFadden, "How Our Female Rowers Ate More and Triumphed," *Newsroom*, August 17, 2021, https://www.newsroom.co.nz/lockerroom/how-our-female-rowers-ate-more-and-triumphed.

66 **women have long struggled:** Samantha L. Moss et al., "Assessment of Energy Availability and Associated Risk Factors in Professional Female Soccer Players," *European Journal of Sport Science* 21, no. 6 (2021): 861–70, https://doi.org/10.1080/17461391.2020.1788647; James C. Morehen et al., "Energy Expenditure of Female International Standard Soccer Players," *Medicine and Science in Sports and Exercise* 54, no. 5 (May 2022): 769–79, https://doi.org/10.1249/MSS.0000000000002850.

66 **people just can't keep up:** Joseph E. Donnelly et al., "Does Increased Exercise or Physical Activity Alter Ad-Libitum Daily Energy Intake or Macronutrient Composition in Healthy Adults? A Systematic Review," *PLOS ONE* 9, no. 1 (January 15, 2014): e83498, https://doi.org/10.1371/journal.pone.0083498.

66 **household and caregiving responsibilities:** Anne Helen Petersen, "'Other Countries Have Social Safety Nets. The U.S. Has Women,'" *Culture Study*, November 11, 2020, https://annehelen.substack.com/p/other-countries-have-social-safety.

67 **there's also often a fraught relationship:** Peiling Kong and Lynne Harris, "The Sporting Body: Body Image and Eating Disorder Symptomatology Among Female Athletes from Leanness Focused and Nonleanness Focused Sports," *The Journal of Psychology Interdisciplinary and Applied* 149, no. 2 (2015): 141–60, https://doi.org/10.1080/00223980.2013.846291.

67 **society's obsession with diet culture:** Michael Hobbes and Aubrey Gordon, "Anti-Fat Bias," November 24, 2020, in *Maintenance Phase*, podcast, https://podcasts.apple.com/us/podcast/anti-fat-bias/id1535408667?i=1000500120510.

67 **professional rock climber Beth Rodden:** Beth Rodden, "Climbing's Send-at-All-Costs Culture Almost Ruined Me," *Outside*, May 2, 2020, https://www.outsideonline.com/health/training-performance/beth-rodden-climbing-body-image/.

68 **relationship between sports and nutrition:** Toni M. Torres-McGehee et al., "Sports Nutrition Knowledge Among Collegiate Athletes, Coaches, Athletic Trainers, and Strength and Conditioning Specialists," *Journal of Athletic Training* 47, no. 2 (2012): 205–11, https://doi.org/10.4085/1062-6050-47.2.205.

69 **At a conference in 2021:** Stellingwerff discussed his informal audit during the 2021 Female Athlete Conference hosted by the Female Athlete Program at Boston Children's Hospital.

70 **"the female athlete triad":** C. L. Otis et al., "American College of Sports Medicine Position Stand: The Female Athlete Triad," *Medicine and Science in Sports and Exercise* 29, no. 5 (May 1997): i–ix, https://doi.org/10.1097/00005768-199705000-00037.

70 **Loucks led a series of seminal studies:** Anne B. Loucks, "Exercise Training in the Normal Female: Effects of Low Energy Availability on Reproductive Function," in *Endocrinology of Physical Activity and Sport*, ed. Naama Constantini and Anthony C. Hackney (New York: Humana Press, 2013), 187–99.

71 **three interrelated conditions:** Elizabeth Joy et al., "2014 Female Athlete Triad Coalition Consensus Statement on Treatment and Return to Play of the Female Athlete Triad," *Current Sports Medicine Reports* 13, no. 4 (July–August 2014): 219–32, https://doi.org/10.1249/JSR.0000000000000077.

71 **"relative energy deficiency in sport" (RED-S):** Margo Mountjoy et al., "The IOC Consensus Statement: Beyond the Female Athlete Triad—Relative Energy Deficiency in Sport (RED-S)," *British Journal of Sports Medicine* 48, no. 7 (2014): 491–97, https://doi.org/10.1136/bjsports-2014-093502.

71 **there's some contention:** Mary Jane De Souza et al., "Misunderstanding the Female Athlete Triad: Refuting the IOC Consensus Statement on Relative Energy Deficiency in Sport (RED-S)," *British Journal of Sports Medicine* 48, no. 20 (2014): 1461–65, https://doi.org/10.1136/bjsports-2014-093958; Nancy I. Williams et al., "Female Athlete Triad and Relative Energy Deficiency in Sport: A Focus on Scientific Rigor," *Exercise and Sport Sciences Reviews* 47, no. 4 (October 2019): 197–205, https://doi.org/10.1249/JES.0000000000000200.

73 **switches to conservation mode:** Lauren M. McCall and Kathryn E. Ackerman, "Endocrine and Metabolic Repercussions of Relative Energy Deficiency in Sport," *Current Opinion in Endocrine and Metabolic Research* 9 (December 2019): 56–65, https://doi.org/10.1016/j.coemr.2019.07.005; F. L. Greenway, "Physiological Adaptations to Weight Loss and Factors Favouring Weight Regain," *International Journal of Obesity* 39, no. 8 (August 2015): 1188–96, https://doi.org/10.1038/ijo.2015.59.

73 **Researchers have observed:** Åsa B. Tornberg et al., "Reduced Neuromuscular Performance in Amenorrheic Elite Endurance Athletes," *Medicine and Science in Sports and Exercise* 49, no. 12 (December 2017): 2478–85, https://doi.org/10.1249/MSS.0000000000001383; Trent Stellingwerff et al., "Overtraining Syndrome (OTS) and Relative Energy Deficiency in Sport (RED-S): Shared Pathways, Symptoms and Complexities," *Sports Medicine* 51, no. 11 (November 2021): 2251–80, https://doi.org/10.1007/s40279-021-01491-0.

73 **risk of bone stress injuries:** Ida A. Heikura et al., "Low Energy Availability Is Difficult to Assess but Outcomes Have Large Impact on Bone Injury Rates in Elite Distance Athletes," *International Journal of Sport Nutrition and Exercise Metabolism* 28, no. 4 (2018): 403–11, https://doi.org/10.1123/ijsnem.2017-0313; Kathryn E. Ackerman et al., "Fractures in Relation to Menstrual Status and Bone Parameters in Young Athletes," *Medicine and Science in Sports and Exercise* 47, no. 8 (August 2015): 1577–86, https://doi.org/10.1249/MSS.0000000000000574; Rayan Ihle and Anne B. Loucks, "Dose-Response Relationships Between Energy Availability and Bone Turnover in Young Exercising Women," *Journal of Bone and Mineral Research* 19, no. 8 (2004): 1231–40, https://doi.org/10.1359/JBMR.040410.

74 **at risk for heart disease:** Paulina Wasserfurth et al., "Reasons for and Consequences of Low Energy Availability in Female and Male Athletes: Social Environment, Adaptations, and Prevention," *Sports Medicine—Open* 6, 44 (September 10, 2020), https://doi.org/10.1186/s40798-020-00275-6.

74 **Experts also warn against:** I. L. Fahrenholtz et al., "Within-Day Energy Deficiency and Reproductive Function in Female Endurance Athletes," *Scandinavian Journal of Medicine and Science in Sports* 28, no. 3 (March 2018): 1139–46, https://doi.org/10.1111/sms.13030.

75 **the Female and Male Athlete Triad:** Michael Fredericson et al., "The Male Athlete Triad—A Consensus Statement from the Female and Male Athlete Triad Coalition. Part II: Diagnosis, Treatment, and Return-to-Play," *Clinical Journal of Sport Medicine* 31, no. 4 (July 2021): 349–66, https://doi.org/10.1097/JSM.0000000000000948.

76 **stems from promising research:** For example, Emily N. C. Manoogian et al., "Time-Restricted Eating for the Prevention and Management of Metabolic Diseases," *Endocrine Reviews* 43, no. 2 (April 2022): 405–36, https://doi.org/10.1210/endrev/bnab027; Mehrdad Alirezaei et al., "Short-Term Fasting Induces Profound Neuronal Autophagy," *Autophagy* 6, no. 6 (August 16, 2010): 702–10, https://doi.org/10.4161/auto.6.6.12376.

76 **it was a toss-up:** T. P. Aird, R. W. Davies, and B. P. Carson, "Effects of Fasted vs. Fed-State Exercise on Performance and Post-Exercise Metabolism: A Systematic Review and Meta-Analysis," *Scandinavian Journal of Medicine and Science in Sports* 28, no. 5 (2018): 1476–93, https://doi.org/10.1111/sms.13054.

76 **when they ate before exercise:** Stephen R. Stannard et al., "Adaptations to Skeletal Muscle with Endurance Exercise Training in the Acutely Fed Versus Overnight-Fasted State," *Journal of Science and Medicine in Sport* 13, no. 4 (July 2010): 465–69, https://doi.org/10.1016/j.jsams.2010.03.002.

78 **As Michael Easter writes:** Michael Easter, "Inside the Rise of Keto: How an Extreme Diet Went Mainstream," *Men's Health*, January 10, 2019, https://www.menshealth.com/nutrition/a25775330/keto-diet-history/.

78 **One of the most rigorous studies:** Louise M. Burke et al., "Crisis of Confidence Averted: Impairment of Exercise Economy and Performance in Elite Race Walkers by Ketogenic Low Carbohydrate, High Fat (LCHF) Diet Is Reproducible," *PLOS ONE* 15, no. 6 (June 4, 2020), https://doi.org/10.1371/journal.pone.0234027.

78 **utilize a greater percentage of fat:** Anthony C. Hackney, Mary Ann McCracken-Compton, and Barbara Ainsworth, "Substrate Responses to Submaximal Exercise in the Midfollicular and Midluteal Phases of the Menstrual Cycle," *International Journal of Sport Nutrition and Exercise Metabolism* 4, no. 3 (1994): 299–308, https://doi.org/10.1123/ijsn.4.3.299; Kealey J. Wohlgemuth et al., "Sex Differences and Considerations for Female Specific Nutritional Strategies: A Narrative Review," *Journal of the International Society of Sports Nutrition* 18, no. 1 (April 1, 2021), https://doi.org/10.1186/s12970-021-00422-8.

78 **Women may also need more carbohydrates:** Sources include interview with Stacy Sims; Megan A. Kuikman et al., "A Review of Nonpharmacological Strategies in the Treatment of Relative Energy Deficiency in Sport," *International Journal of Sport Nutrition and Exercise Metabolism* 31, no. 3 (January 19, 2021): 268–75, https://doi.org/10.1123/ijsnem.2020-0211.

79 **bump up cortisol levels:** Radhika V. Seimon et al., "Effects of Energy Restriction on Activity of the Hypothalamo-Pituitary-Adrenal Axis in Obese Humans and Rodents: Implications for Diet-Induced Changes in Body Composition," *Hormone Molecular Biology and Clinical Investigation* 15, no. 2 (2013): 71–80, https://doi.org/10.1515/hmbci-2013-0038.

80 **"hierarchy of nutritional needs":** Bryan Holtzman and Kathryn E. Ackerman, "Recommendations and Nutritional Considerations for Female Athletes: Health and Performance," *Sports Medicine* 51, no. 9 Suppl. (2021): 43–57, https://doi.org/10.1007/s40279-021-01508-8.

80 **junior elite swimmers:** Jaci L. VanHeest et al., "Ovarian Suppression Impairs Sport Performance in Junior Elite Female Swimmers," *Medicine and Science in Sports and Exercise* 46, no. 1 (January 2014): 156–66, https://doi.org/10.1249/MSS.0b013e3182a32b72.

81 **researchers from Penn State University:** Mary Jane De Souza et al., "Randomised Controlled Trial of the Effects of Increased Energy Intake on Menstrual Recovery in Exercising Women with Menstrual Disturbances: The 'REFUEL' Study," *Human Reproduction* 36, no. 8 (August 2021): 2285–97, https://doi.org/10.1093/humrep/deab149.

81 **researchers found that middle-distance runners:** R. C. Deutz et al., "Relationship Between Energy Deficits and Body Composition in Elite Female Gymnasts and Runners," *Medicine and Science in Sports and Exercise* 32, no. 3 (March 2000): 659–68, https://doi.org/10.1097/00005768-200003000-00017.

5. The Long Game

84 **Sarah Thomas just wanted to know:** Sources for Sarah Thomas's section include interviews with Thomas, Elaine Howley, and Evan Morrison, co-founder of the Marathon Swimmers Federation; Elaine Howley, "The Other Side," *Outdoor Swimmer*, October 2019, 32–37;

Jon Washer, "The Other Side—A Documentary Film About Open Water Swimming," You-Tube, August 15, 2019, https://youtu.be/cNCHBHKi38g. A database of Thomas's swims can be found here: "Sarah Thomas," LongSwims Database, https://longswims.com/p/sarah -thomas/.

86 **Since 1875:** "Solo Swim Statistics," Channel Swimming and Piloting Federation, http:// cspf.co.uk/solo-swims-statistics.

86 **women swimmers like Gertrude Ederle:** "Gertrude Ederle becomes first woman to swim English Channel," *History* (blog), April 16, 2021, https://www.history.com/this-day -in-history/gertrude-ederle-becomes-first-woman-to-swim-english-channel.

89 **physiology professors Brian Whipp:** B. J. Whipp and S. A. Ward, "Will Women Soon Outrun Men?," *Nature* 355, no. 6355 (January 2, 1992): 25, https://doi.org/10.1038/355025a0.

89 **fundamental differences in anatomy and physiology:** Nicholas B. Tiller et al., "Do Sex Differences in Physiology Confer a Female Advantage in Ultra-Endurance Sport?," *Sports Medicine* 51, no. 5 (May 2021): 895–915, https://doi.org/10.1007/s40279-020-01417-2.

90 **across athletic disciplines, women's records:** Sandra Hunter shared a slide with me from her presentation "Sex Differences in Human Performance: Power of Big Data" given on November 11, 2020. In 2019, the differential between the men's and women's records in running events (100 meters, 1500 meters, 10,000 meters, and marathon) was between 9.5 and 11.7 percent. The gap was smaller for long-distance swimming events and bigger for power-based events like the high jump and long jump.

90 **researchers examined more than five million results:** Paul Ronto, "The State of Ultra Running 2020," *RunRepeat,* September 21, 2021, https://runrepeat.com/state-of-ultra-running.

91 **a case of sampling bias:** Sandra K. Hunter and Alyssa A. Stevens, "Sex Differences in Marathon Running with Advanced Age: Physiology or Participation?," *Medicine and Science in Sports and Exercise* 45, no. 1 (January 2013): 148–56, https://doi.org/10.1249/MSS.0b013 e31826900f6; Jonathon Senefeld, Carolyn Smith, and Sandra K. Hunter, "Sex Differences in Participation, Performance, and Age of Ultramarathon Runners," *International Journal of Sports Physiology and Performance* 11, no. 7 (January 2016): 635–42, https://doi.org/10.1123 /ijspp.2015-0418.

91 **2021 Western States Endurance Run:** iRunFar (@iRunFar), "The race has confirmed this is the first time in Western States 100 history that three women finished in the overall top ten, #WS100," Twitter, June 27, 2021, https://twitter.com/iRunFar/status/14090 36313144090626; iRunFar (@iRunFar), "This year's Western States 100 had a 66% finishing rate . . . ," Twitter, June 26, 2021, https://twitter.com/iRunFar/status/1409236864842665985.

92 **ultrarunner Camille Herron:** Sources for the Camille Herron section include an interview with Herron; Amby Burfoot, "Camille Herron Crushes the Ultra Competition with a Smile," *Runner's World,* January 16, 2019, https://www.runnersworld.com/runners-stories /a25916911/camille-herron-on-ultrarunning-success/; Taylor Dutch, "Camille Herron Never Accepts 'No' for an Answer," *Runner's World,* December 30, 2019, https://www.runner sworld.com/runners-stories/a30361143/camille-herron-ultrarunning-success/; Amby Burfoot, "Camille Herron Will Run Her 100,000th Mile This Week," *Outside,* April 6, 2022, https://www.outsideonline.com/health/running/culture-running/people/camille-herron -100k-mile-club/.

94 **one of three factors:** Michael J. Joyner et al., "Physiology and Fast Marathons," *Journal of Applied Physiology* 128, no. 4 (April 2020): 1065–68, https://doi.org/10.1152/japplphysiol .00793.2019.

95 **greater distribution of slow-twitch:** Sandra K. Hunter, "Sex Differences in Human Fatigability: Mechanisms and Insight to Physiological Responses," *Acta Physiologica* 210, no. 4 (April 2014): 768–89, https://doi.org/10.1111/apha.12234; Sandra K. Hunter, "The Relevance of Sex Differences in Performance Fatigability," *Medicine and Science in Sports and Exercise* 48, no. 11 (November 2016): 2247–56, https://doi.org/10.1249/MSS.0000000000000928.

95 **women tire less quickly:** Sandra K. Hunter et al., "Men Are More Fatigable than Strength-Matched Women When Performing Intermittent Submaximal Contractions," *Journal of Applied Physiology* 96, no. 6 (June 2004): 2125–32, https://doi.org/10.1152/japplphysiol .01342.2003.

95 **greater muscle mass and muscle fiber diameter:** Hunter, "Sex Differences in Human Fatigability," 768–89; William S. Barnes, "The Relationship Between Maximum Isometric

Strength and Intramuscular Circulatory Occlusion," *Ergonomics* 23, no. 4 (April 1980): 351–57, https://doi.org/10.1080/00140138008924748.

96 **Canadian and French researchers:** John Temesi et al., "Are Females More Resistant to Extreme Neuromuscular Fatigue?," *Medicine and Science in Sports and Exercise* 47, no. 7 (July 2015): 1372–82, https://doi.org/10.1249/MSS.0000000000000540.

97 **women are more metabolically nimble:** Jennifer Wismann and Darryn Willoughby, "Gender Differences in Carbohydrate Metabolism and Carbohydrate Loading," *Journal of the International Society of Sports Nutrition* 3, no. 1 (2006): 28–34, https://doi.org/10.1186 /1550-2783-3-1-28; Hannah N. Willett, Kristen J. Koltun, and Anthony C. Hackney, "Influence of Menstrual Cycle Estradiol-β-17 Fluctuations on Energy Substrate Utilization-Oxidation During Aerobic, Endurance Exercise," *International Journal of Environmental Research and Public Health* 18, no. 13 (July 5, 2021): 7209, https://doi.org/10.3390/ijerph 18137209; Amy C. Maher et al., "Women Have Higher Protein Content of β-Oxidation Enzymes in Skeletal Muscle Than Men," *PLOS ONE* 5, no. 8 (August 6, 2010), https://doi .org/10.1371/journal.pone.0012025.

97 **submaximal efforts typical of endurance events:** Part of the difference in fat and carbohydrate utilization between women and men could be due to the fact that women store less carbohydrate compared to men and therefore have less carbohydrate on hand available to use. Additionally, this metabolic advantage is specific to submaximal exercise. During high-intensity exercise, men may have an advantage because energy required to fuel high-intensity activity is derived from carbohydrates.

97 **top two reasons athletes drop out:** Martin D. Hoffman and Kevin Fogard, "Factors Related to Successful Completion of a 161-Km Ultramarathon," *International Journal of Sports Physiology and Performance* 6, no. 1 (2011): 25–37, https://doi.org/10.1123/ijspp.6.1.25.

97 **an advantage in marathon swimming:** Beat Knechtle et al., "Sex Differences in Swimming Disciplines—Can Women Outperform Men in Swimming?," *International Journal of Environmental Research and Public Health* 17, no. 10 (May 2020): 3651, https://doi.org /10.3390/ijerph17103651; Beat Knechtle, Thomas Rosemann, and Christoph Alexander Rüst, "Women Cross the 'Catalina Channel' Faster Than Men," *SpringerPlus* 4, 32 (July 8, 2015), https://doi.org/10.1186/s40064-015-1086-4; Beat Knechtle et al., "Women Outperform Men in Ultradistance Swimming: The Manhattan Island Marathon Swim from 1983 to 2013," *International Journal of Sports Physiology and Performance* 9, no. 6 (November 2014): 913–24, https://doi.org/10.1123/ijspp.2013-0375.

99 **Courtney Dauwalter has seen:** Sources for the Courtney Dauwalter section include an interview with Dauwalter; Rebecca Byerly, "The Woman Who Outruns the Men, 200 Miles at a Time," *The New York Times*, December 5, 2018, https://www.nytimes.com/2018/12/05 /sports/courtney-dauwalter-200-mile-race.html; Brian Metzler, "Courtney Dauwalter Breaks UTMB Record to Win," *Trail Runner Magazine*, August 30, 2021, https://www.trailrunner mag.com/people/courtney-dauwalters-record-breaking-utmb.

102 **"the psycho-biological model of fatigue":** Samuele Marcora, "Psychobiology of Fatigue During Endurance Exercise," in *Endurance Performance in Sport*, ed. Carla Meijen (London: Routledge, 2019), 15–34, https://doi.org/10.4324/9781315167312-2.

102 **surveyed 344 women ultrarunners:** Rhonna Z. Krouse et al., "Motivation, Goal Orientation, Coaching, and Training Habits of Women Ultrarunners," *The Journal of Strength and Conditioning Research* 25, no. 10 (October 2011): 2835–42, https://doi.org/10.1519/JSC.0b013 e318204caa0.

104 **men tend to be more overconfident:** Calvin Hubble and Jinger Zhao, "Gender Differences in Marathon Pacing and Performance Prediction," *Journal of Sports Analytics* 2, no. 1 (2016): 19–36, https://doi.org/10.3233/JSA-150008; Robert O. Deaner et al., "Men Are More Likely Than Women to Slow in the Marathon," *Medicine and Science in Sports and Exercise* 47, no. 3 (March 2015): 607–16, https://doi.org/10.1249/MSS.0000000000000432; Andrew Renfree, Everton Crivoi do Carmo, and Louise Martin, "The Influence of Performance Level, Age and Gender on Pacing Strategy During a 100-Km Ultramarathon," *European Journal of Sport Science* 16, no. 4 (2016): 409–15, https://doi.org/10.1080/17461391.2015.1041061.

104 **2008 to 2013 BOLDERBoulder event:** Robert O. Deaner et al., "Fast Men Slow More Than Fast Women in a 10 Kilometer Road Race," *PeerJ* 4 (2016), https://doi.org/10.7717/peerj .2235.

6. The Dreaded Female Body

107 **Briana Scurry doesn't recall:** In addition to an interview with Briana Scurry, for more on Scurry's experience with injury and post-concussion, see Caitlin Dewey, "Her Biggest Save," *The Washington Post*, November 2, 2013, https://www.washingtonpost.com/sf/national/2013 /11/02/her-biggest-save.

109 **NCAA Injury Surveillance System:** Tracey Covassin, C. Buz Swanik, and Michael L. Sachs, "Epidemiological Considerations of Concussions Among Intercollegiate Athletes," *Applied Neuropsychology* 10, no. 1 (2003): 12–22, https://doi.org/10.1207/S15324826AN1001_3.

110 **when Covassin and her colleagues:** Scott L. Zuckerman et al., "Epidemiology of Sports-Related Concussion in NCAA Athletes from 2009–2010 to 2013–2014: Incidence, Recurrence, and Mechanisms," *The American Journal of Sports Medicine* 43, no. 11 (November 2015): 2654–62, https://doi.org/10.1177/0363546515599634.

110 **women seemed to hurt their knees:** Elizabeth A. Arendt, Julie Agel, and Randall Dick, "Anterior Cruciate Ligament Injury Patterns Among Collegiate Men and Women," *Journal of Athletic Training* 34, no. 2 (1999): 86–92; Timothy E. Hewett et al., "Understanding and Preventing ACL Injuries: Current Biomechanical and Epidemiologic Considerations— Update 2010," *North American Journal of Sports Physical Therapy* 5, no. 4 (December 2010): 234–51.

111 **they also experienced worse outcomes:** J. Larruskain et al., "A Comparison of Injuries in Elite Male and Female Football Players: A Five-Season Prospective Study," *Scandinavian Journal of Medicine and Science in Sports* 28, no. 1 (2018): 237–45, https://doi.org/10.1111 /sms.12860; Mark V. Paterno et al., "Incidence of Second ACL Injuries 2 Years After Primary ACL Reconstruction and Return to Sport," *The American Journal of Sports Medicine* 42, no. 7 (July 2014): 1567–73, https://doi.org/10.1177/0363546514530088; Kathryn L. Van Pelt et al., "Detailed Description of Division I Ice Hockey Concussions: Findings from the NCAA and Department of Defense CARE Consortium," *Journal of Sport and Health Science* 10, no. 2 (March 2021): 162–71, https://doi.org/10.1016/j.jshs.2021.01.004.

112 **women are still left out:** Paul McCrory et al., "Consensus Statement on Concussion in Sport—the 5th International Conference on Concussion in Sport Held in Berlin, October 2016," *British Journal of Sports Medicine* 51, no. 11 (2017): 838–47, https://doi.org/10.1136 /bjsports-2017-097699; Thomas L. Sanders, et al., "Incidence of Anterior Cruciate Ligament Tears and Reconstruction: A 21-Year Population-Based Study," *The American Journal of Sports Medicine* 44, no. 6 (June 2016): 1502–7, https://doi.org/10.1177/0363546516629944.

115 **recruit the quadriceps muscles:** Timothy E. Hewett et al., "Understanding and Preventing ACL Injuries," 234–51.

116 **Injury rates between girls and boys:** The incidence of ACL injury rose the fastest in girls ages fourteen to seventeen. See Rick P. Csintalan, Maria C. S. Inacio, and Tadashi T. Funahashi, "Incidence Rate of Anterior Cruciate Ligament Reconstructions," *The Permanente Journal* 12, no. 3 (2008): 17–21, https://doi.org/10.7812/tpp/07-140.

116 **women experience higher acceleration:** Ryan T. Tierney et al., "Gender Differences in Head-Neck Segment Dynamic Stabilization During Head Acceleration," *Medicine and Science in Sports and Exercise* 37, no. 2 (February 2005): 272–79, https://doi.org/10.1249 /01.MSS.0000152734.47516.AA.

117 **they created "mini-brains":** Jean-Pierre Dollé et al., "Newfound Sex Differences in Axonal Structure Underlie Differential Outcomes from In Vitro Traumatic Axonal Injury," *Experimental Neurology* 300 (February 2018): 121–34, https://doi.org/10.1016/j.expneurol .2017.11.001.

118 **Estrogen has received:** Nkechinyere Chidi-Ogbolu and Keith Baar, "Effect of Estrogen on Musculoskeletal Performance and Injury Risk," *Frontiers in Physiology* 9 (2019), https://doi .org/10.3389/fphys.2018.01834; Cassandra A. Lee et al., "Estrogen Inhibits Lysyl Oxidase and Decreases Mechanical Function in Engineered Ligaments," *Journal of Applied Physiology* 118, no. 10 (May 2015): 1250–57, https://doi.org/10.1152/japplphysiol.00823.2014.

118 **knees become more lax:** Some studies examining knee laxity include Sang-Kyoon Park et al., "Changing Hormone Levels During the Menstrual Cycle Affect Knee Laxity and Stiffness in Healthy Female Subjects," *The American Journal of Sports Medicine* 37, no. 3 (March 2009): 588–98, https://doi.org/10.1177/0363546508326713; Haneul Lee et al., "Anterior

Cruciate Ligament Elasticity and Force for Flexion During the Menstrual Cycle," *Medical Science Monitor* 19 (2013): 1080–88, https://doi.org/10.12659/MSM.889393.

118 **lower risk of an Achilles tendon rupture:** Karsten Hollander et al., "Sex-Specific Differences in Running Injuries: A Systematic Review with Meta-Analysis and Meta-Regression," *Sports Medicine* 51, no. 5 (May 2021): 1011–39, https://doi.org/10.1007/s40279-020-01412-7.

119 **Women may also be less susceptible:** Pascal Edouard, Pedro Branco, and Juan-Manuel Alonso, "Muscle Injury Is the Principal Injury Type and Hamstring Muscle Injury Is the First Injury Diagnosis During Top-Level International Athletics Championships between 2007 and 2015," *British Journal of Sports Medicine* 50, no. 10 (2016): 619–30, https://doi.org/10.1136/bjsports-2015-095559.

119 **emotional trauma can linger:** Julie P. Burland et al., "Decision to Return to Sport After Anterior Cruciate Ligament Reconstruction, Part I: A Qualitative Investigation of Psychosocial Factors," *Journal of Athletic Training* 53, no. 5 (May 2018): 452–63, https://doi.org/10.4085/1062-6050-313-16; Ajay S. Padaki et al., "Prevalence of Posttraumatic Stress Disorder Symptoms Among Young Athletes After Anterior Cruciate Ligament Rupture," *Orthopaedic Journal of Sports Medicine* 6, no. 7 (July 2018), https://doi.org/10.1177/2325967118787159.

120 **Among recreational women athletes:** Elaine Mullally, "Non-contact Knee Injury Prevention in Adult Recreational Female Netball Players: A Review of Current Understanding with Future Research," presentation, Women in Sport and Exercise (WiSE) Conference, virtual, April 20, 2021.

120 **thirteen times more likely:** Mark V. Paterno et al., "Self-Reported Fear Predicts Functional Performance and Second ACL Injury After ACL Reconstruction and Return to Sport: A Pilot Study," *Sports Health* 10, no. 3 (May 2018): 228–33, https://doi.org/10.1177/1941738117745806.

120 ***British Journal of Sports Medicine*:** Joanne L. Parsons, Stephanie E. Coen, and Sheree Bekker, "Anterior Cruciate Ligament Injury: Towards a Gendered Environmental Approach," *British Journal of Sports Medicine* 55, no. 17 (September 2021): 984–90, https://doi.org/10.1136/bjsports-2020-103173.

120 **Australian Football League Women's (AFLW):** Aaron Fox et al., "Anterior Cruciate Ligament Injuries in Australian Football: Should Women and Girls Be Playing? You're Asking the Wrong Question," *BMJ Open Sport and Exercise Medicine* 6, no. 1 April, https://doi.org/10.1136/bmjsem-2020-000778.

121 **To test this hypothesis:** Karl F. Orishimo et al., "Comparison of Landing Biomechanics Between Male and Female Professional Dancers," *The American Journal of Sports Medicine* 37, no. 11 (November 2009): 2187–93, https://doi.org/10.1177/0363546509339365; Karl F. Orishimo et al., "Comparison of Landing Biomechanics Between Male and Female Dancers and Athletes. Part 1: Influence of Sex on Risk of Anterior Cruciate Ligament Injury," *The American Journal of Sports Medicine* 42, no. 5 (May 2014): 1082–88, https://doi.org/10.1177/0363546514523928.

122 **to visibly demonstrate their manhood:** Stephanie E. Coen, Mark W. Rosenberg, and Joyce Davidson, "'It's Gym, like g-y-m Not J-i-m': Exploring the Role of Place in the Gendering of Physical Activity," *Social Science and Medicine* 196 (January 2018): 29–36, https://doi.org/10.1016/j.socscimed.2017.10.036.

122 **survey of high school varsity coaches:** Monica L. Reynolds et al., "An Examination of Current Practices and Gender Differences in Strength and Conditioning in a Sample of Varsity High School Athletic Programs," *The Journal of Strength and Conditioning Research* 26, no. 1 (January 2012): 174–83, https://doi.org/10.1519/JSC.0b013e31821852b7.

123 **Women fencers, in particular:** Kamali Thompson et al., "Lower Extremity Injuries in U.S. National Fencing Team Members and U.S. Fencing Olympians," *The Physician and Sportsmedicine* 50, no. 3 (2022): 212–17, https://doi.org/10.1080/00913847.2021.1895693.

124 **Master was part:** Natasha Desai et al., "Factors Affecting Recovery Trajectories in Pediatric Female Concussion," *Clinical Journal of Sport Medicine* 29, no. 5 (September 2019): 361–67, https://doi.org/10.1097/JSM.0000000000000646.

124 **data on more than one thousand concussions:** Christina L. Master et al., "Differences in Sport-Related Concussion for Female and Male Athletes in Comparable Collegiate

Sports: A Study from the NCAA-DoD Concussion Assessment, Research and Education (CARE) Consortium," *British Journal of Sports Medicine* 55, no. 24 (December 2021): 1387–94, https://doi.org/10.1136/bjsports-2020-103316.

125 **The Football Association:** Fiona Thomas, "FA 'Treating Women's Footballers as Second-Class Citizens' Over Concussion Risk," *The Telegraph*, April 2, 2021, https://www.telegraph.co.uk/football/2021/04/02/exclusivefa-accused-treating-womens-footballers-second-class/.

126 **program for the Hockeyroos:** Gillian Weir et al., "A 2-Yr Biomechanically Informed ACL Injury Prevention Training Intervention in Female Field Hockey Players," *Translational Journal of the American College of Sports Medicine* 4, no. 19 (October 2019): 206–14, https://journals.lww.com/acsm-tj/fulltext/2019/10010/a_2_yr_biomechanically_informed_acl_injury.2.aspx.

127 **Overall, ACL injury prevention programs:** Kate E. Webster and Timothy E. Hewett, "Meta-Analysis of Meta-Analyses of Anterior Cruciate Ligament Injury Reduction Training Programs," *Journal of Orthopaedic Research* 36, no. 10 (October 2018): 2696–708, https://doi.org/10.1002/jor.24043.

129 **according to Ackerman:** "Supporting Performance and Health of Female Athletes," interviewed by Scott L. Delp, launch event for the Wu Tsai Human Performance Alliance, video, 3:30, August 2, 2021, https://youtu.be/nOnbsyM0jTY. Panelists include Kathryn Ackerman, NiCole Keith, and Emily Kraus.

7. Bounce Control

131 **for supporting breasts during physical activity:** Melissa Pandika, "Bra History: How a War Shortage Reshaped Modern Shapewear," NPR, August 5, 2014, https://www.npr.org/2014/08/05/337860700/bra-history-how-a-war-shortage-reshaped-modern-shapewear.

132 **Among those to take up running:** In addition to interviews with Lisa Lindahl and Hinda Miller, sources include Lisa Z. Lindahl, *Unleash the Girls: The Untold Story of the Invention of the Sports Bra and How It Changed the World (and Me)*, (EZL Enterprises, 2019); Allison Keyes, "How the First Sports Bra Got Its Stabilizing Start," *Smithsonian Magazine*, March 18, 2020, https://www.smithsonianmag.com/smithsonian-institution/how-first-sports-bra-got-stabilizing-start-180974427/.

134 **sports bra market:** WinterGreen Research, "Sports Bra: Market Shares, Strategies, and Forecasts, Worldwide, 2020 to 2026," MarketResearch.com, January 2020, https://www.marketresearch.com/Wintergreen-Research-v739/Sports-Bra-Shares-Strategies-Forecasts-12924218/.

135 **breasts are made of soft tissue:** Jenny Burbage, "Breasts," presentation, The Science Behind Bras and Breasts Workshop, September 24, 2020; Deirdre E. McGhee and Julie R. Steele, "Breast Biomechanics: What Do We Really Know?," *Physiology* 35, no. 2 (March 2020): 144–56, https://doi.org/10.1152/physiol.00024.2019; Kelly-Ann Page and Julie R. Steele, "Breast Motion and Sports Brassiere Design," *Sports Medicine* 27, no. 4 (April 1999): 205–11, https://doi.org/10.2165/00007256-199927040-00001.

135 **How much or how fast breasts move:** Brogan Jones, "Breast Biomechanics Research," presentation, The Science Behind Bras and Breasts Workshop, September 24, 2020; Joanna C. Scurr, Jennifer L. White, and Wendy Hedger, "Supported and Unsupported Breast Displacement in Three Dimensions Across Treadmill Activity Levels," *Journal of Sports Sciences* 29, no. 1 (2011): 55–61, https://doi.org/10.1080/02640414.2010.521944; Claire Bridgman et al., "Three-Dimensional Kinematics of the Breast During a Two-Step Star Jump," *Journal of Applied Biomechanics* 26, no. 4 (January 2010): 465–72, https://doi.org/10.1123/jab.26.4.465.

135 **It can be uncomfortable:** Some studies include Nicola Brown, Jennifer White, Amanda Brasher, and Joanna Scurr, "The Experience of Breast Pain (Mastalgia) in Female Runners of the 2012 London Marathon and Its Effect on Exercise Behaviour," *British Journal of Sports Medicine* 48, no. 4 (February 2014): 320–25, https://doi.org/10.1136/bjsports-2013-092175; Brooke R. Brisbine, Julie R. Steele, Elissa J. Phillips, and Deirdre E. McGhee, "Breast Pain Affects the Performance of Elite Female Athletes," *Journal of Sports Sciences* 38, no. 5 (2020): 528–33, https://doi.org/10.1080/02640414.2020.1712016.

137 **Former professional runner Lauren Fleshman:** Lauren Fleshman (@fleshmanflyer), "After 26 years of running, most at a high performance level, I developed some skills . . . ," Instagram photo, January 28, 2021, https://www.instagram.com/p/CKmisnilwat/?igshid= 102m68jwdmeom.

137 **truth to that perception:** J. L. White, J. C. Scurr, and N. A. Smith, "The Effect of Breast Support on Kinetics During Overground Running Performance," *Ergonomics* 52, no. 4 (April 2009): 492–98, https://doi.org/10.1080/00140130802707907; Alexandra Milligan, Chris Mills, Jo Corbett, and Joanna Scurr, "The Influence of Breast Support on Torso, Pelvis and Arm Kinematics During a Five Kilometer Treadmill Run," *Human Movement Science* 42 (August 2015): 246–60, https://doi.org/10.1016/j.humov.2015.05.008; Hailey B. Fong and Douglas W. Powell, "Greater Breast Support Is Associated with Reduced Oxygen Consumption and Greater Running Economy During a Treadmill Running Task," *Frontiers in Sports and Active Living* 4 (June 14, 2022), https://doi.org/10.3389/fspor.2022.902276.

137 **4 centimeters shorter:** Jenny White, Joanna Scurr, and Chris Mills, "Breast Support Implications for Female Recreational Athletes During Steady-State Running," *ISBS— Conference Proceedings Archive*, September 1, 2013, https://ojs.ub.uni-konstanz.de/cpa/art icle/view/5600.

138 ***Outside* survey about sports bras:** Ariella Gintzler, "We're in the Middle of a Sports-Bra Revolution," *Outside*, March 5, 2020, https://www.outsideonline.com/outdoor-gear/clothing -apparel/sports-bra-design-revolution/.

139 **Sports bra sizing:** Jenny Burbage, "Bra Fit and Education," presentation, The Science Behind Bras and Breasts Workshop, September 24, 2020.

140 **breasts can keep people from being active:** Nicola Brown, Jenny Burbage, and Joanna Wakefield-Scurr, "Sports Bra Use, Preferences and Fit Issues Among Exercising Females in the US, the UK and China," *Journal of Fashion Marketing and Management* 25, no. 3 (2021): 511–27, https://doi.org/10.1108/JFMM-05-2020-0084; Emma Burnett, Jenny White, and Joanna Scurr, "The Influence of the Breast on Physical Activity Participation in Females," *Journal of Physical Activity and Health* 12, no. 4 (April 2015): 588–94, https://doi.org/10 .1123/jpah.2013-0236.

141 **hard to study breasts in motion:** I rely on interviews with Jenny Burbage, Brogan Jones, and LaJean Lawson for much of my discussion of breast biomechanics and studies such as Joanna Scurr, Jennifer White, and Wendy Hedger, "Breast Displacement in Three Dimensions During the Walking and Running Gait Cycles," *Journal of Applied Biomechanics* 25, no. 4 (January 2009): 322–29, https://doi.org/10.1123/jab.25.4.322.

142 **Lawson started investigating breast biomechanics:** LaJean Lawson, "A Comparison of Eight Selected Sports Bras: Biomechanical Support, Overall Comfort Ratings and Overall Support Ratings," master's thesis, Utah State University, 1985, https://digitalcommons.usu .edu/etd/4039.

145 **the global athleisure market:** "Athleisure Market Size to Expand at 8.1% CAGR by 2025, Owing to Rising Adaption of Fashionable and Stylish Athleisure Wearing in Offices and Work Places," *Bloomberg*, April 7, 2021, https://www.bloomberg.com/press-releases/2021 -04-07/athleisure-market-size-to-expand-at-8-1-cagr-by-2025-owing-to-rising-adaption -of-fashionable-stylish-athleisure-wearing-in.

147 **sizes ranging from 30A:** Brooks does go up to a G cup but only to a size 40 underband.

148 **do a better job reducing breast movement:** Michelle Norris et al., "How the Character-istics of Sports Bras Affect Their Performance," *Ergonomics* 64, no. 3 (2021): 410–25, https:// doi.org/10.1080/00140139.2020.1829090.

148 **DD cup size on average:** "The Average American Bra Size is Now a 34 DD," *Racked*, July 22, 2013, https://www.racked.com/2013/7/22/7659335/bra-size-increase.

150 **a good running bra:** Deborah Risius et al., "Multiplanar Breast Kinematics During Differ-ent Exercise Modalities," *European Journal of Sport Science* 15, no. 2 (2015): 111–17, https:// doi.org/10.1080/17461391.2014.928914.

150 **Team GB (Great Britain):** "Innovative Sports Bras Aim to Boost Health and Perform-ance," *English Institute of Sport* (blog), March 8, 2021, https://eis2win.co.uk/article/innova tive-sports-bras-aim-to-boost-health-and-performance/; Joanna Wakefield-Scurr, Amy San-chez, and Melissa Jones, "A Multi-Stage Intervention Assessing, Advising and Customising

Sports Bras for Elite Female British Athletes," *Research in Sports Medicine*, February 14, 2022, https://doi.org/10.1080/15438627.2022.2038162.

8. Beyond Shrink It and Pink It

152 **Charlotte Dod's nickname:** Helen Lewis, "In Search of the First Female Sports Super-star," *The Atlantic*, September 13, 2020, https://www.theatlantic.com/magazine/archive/2020/10/was-charlotte-dod-the-greatest-athlete-ever/615490/.

155 **the women's sportswear space:** Cara Salpini, "Game-changers: Have Women Reshaped the Sportswear Market?," *Retail Dive*, September 3, 2019, https://www.retaildive.com/news/game-changers-have-women-reshaped-the-sportswear-market/561607/.

157 **Megan Rapinoe said:** Kathleen McNamee, "Women's Sports Combatting a Surprising Barrier to Entry: Gear Designed with Men in Mind," *ESPN*, March 12, 2021, https://www.espn.com/soccer/blog-espn-fc-united/story/4332912/womens-sports-combatting-a-surprising-barrier-to-entry-gear-designed-with-men-in-mind.

157 **When the writer Latria Graham:** Latria Graham (@mslatriagraham), "I wish I felt better about this gorgeous photo . . . ," Instagram photo, February 4, 2021, https://www.instagram.com/p/CK300L2Ajbu/?igshid=1g05g29t4z9j8.

158 **women are projected to account for:** "Wise Up to Women," *Nielsen*, March, 2, 2020, https://www.nielsen.com/us/en/insights/article/2020/wise-up-to-women/.

159 **Eve O'Neill told *Medium*:** Hannah Weinberger, "Just What Is 'Women's Specific Gear' Anyways? And Do You Need It?," *Medium*, April 10, 2018, https://medium.com/s/story/no-one-really-knows-what-womens-specific-gear-is-and-that-s-a-problem-49c1d536d714.

159 **women tend to feel colder:** L. Schellen et al., "The Influence of Local Effects on Thermal Sensation Under Non-Uniform Environmental Conditions—Gender Differences in Ther-mophysiology, Thermal Comfort and Productivity During Convective and Radiant Cooling," *Physiology and Behavior* 107, no. 2 (September 10, 2012): 252–61, https://doi.org/10.1016/j.physbeh.2012.07.008.

159 **The bike saddle:** Andy Pruitt provided most of the details about the bike saddles design and development process.

161 **she wasn't the only cyclist:** James J. Potter et al., "Gender Differences in Bicycle Saddle Pressure Distribution During Seated Cycling," *Medicine and Science in Sports and Exercise* 40, no. 6 (June 2008): 1126–34, https://doi.org/10.1249/MSS.0b013e3181666eea.

162 **breasts are vulnerable:** Brooke R. Brisbine et al., "The Occurrence, Causes and Perceived Performance Effects of Breast Injuries in Elite Female Athletes," *Journal of Sports Science and Medicine* 18, no. 3 (September 2019): 569–76.

162 **women's feet are V-shaped:** Gangming Luo et al., "Comparison of Male and Female Foot Shape," *Journal of the American Podiatric Medical Association* 99, no. 5 (2009): 383–90, https://doi.org/10.7547/0990383.

163 **Geoff Hollister's book:** Geoff Hollister, *Out of Nowhere: The Inside Story of How Nike Marketed the Culture of Running* (United Kingdom: Meyer & Meyer Sport, 2008), 101.

165 **It's been hypothesized:** Many studies have tested this theory, but the majority suggest that a higher Q angle doesn't contribute to knee stress and injuries. See Tahani A. Alahmad, Philip Kearney, and Roisin Cahalan, "Injury in Elite Women's Soccer: A Systematic Review," *The Physician and Sportsmedicine* 48, no. 3 (2020): 259–65, https://doi.org/10.1080/00913847.2020.1720548; Bryan C. Heiderscheit, Joseph Hamill, and Graham E. Caldwell, "In-fluence of Q-Angle on Lower-Extremity Running Kinematics," *Journal of Orthopaedic and Sports Physical Therapy* 30, no. 5 (May 2000): 271–78, https://doi.org/10.2519/jospt.2000.30.5.271.

165 **the effect of running shoes:** D. Casey Kerrigan et al., "The Effect of Running Shoes on Lower Extremity Joint Torques," *PM&R* 1, no. 12 (December 2009): 1058–63, https://doi.org/10.1016/j.pmrj.2009.09.011.

165 **when women and men walked barefoot:** D. C. Kerrigan et al., "Knee Joint Torques: A Comparison Between Women and Men During Barefoot Walking," *Archives of Physical Medicine and Rehabilitation* 81, no. 9 (September 2000): 1162–65, https://doi.org/10.1053/apmr.2000.7172.

166 **The plus-size market:** P. Smith, "U.S. Women's Plus-Size Apparel Market—Statistics and Facts," *Statista*, January 12, 2022, https://www.statista.com/topics/4834/women-s-plus-size-apparel-market-in-the-us.

170 **ill-fitting boots are problematic:** Katharina Althoff and Ewald M. Hennig, "Criteria for Gender-Specific Soccer Shoe Development," *Footwear Science* 6, no. 2 (2014): 89–96, https://doi.org/10.1080/19424280.2014.890671.

173 **"research into the Retül database":** Rita Jett, Samir Chabra, and Todd Carver, "When to Share Product Platforms: An Anthropometric Review," whitepaper, Specialized, https://specialized.picturepark.com/Go/drxAWjId/D/59574/1.

174 **Lintilhac told *GearJunkie*:** Berne Broudy, "Do Women Need Women's Skis?," *GearJunkie*, February 13, 2020, https://gearjunkie.com/winter/skiing/do-women-need-womens-skis.

9. The Phenom Years

179 **In gymnastics, for instance:** Dvora Meyers, "Time for the End of the Teen Gymnast," *FiveThirtyEight*, July 27, 2021, https://fivethirtyeight.com/features/gymnasts-age-olympics/.

180 **"10,000-hour rule":** Malcolm Gladwell, *Outliers* (New York: Little, Brown and Company, 2008), 35–68.

180 **1993 study of violinists:** K. Anders Ericsson, R. T. Krampe, and Clemens Tesch-Römer, "The Role of Deliberate Practice in the Acquisition of Expert Performance," *Psychological Review* 100, no. 3 (1993): 363–406, https://doi.org/10.1037/0033-295X.100.3.363.

180 **study's authors, K. Anders Ericsson:** K. Anders Ericsson, "Training History, Deliberate Practice and Elite Sports Performance: An Analysis in Response to Tucker and Collins Review—What Makes Champions?," *British Journal of Sports Medicine* 47, no. 9 (June 2013): 533–35, https://doi.org/10.1136/bjsports-2012-091767.

180 **an entire ecosystem has emerged:** David R. Bell et al., "The Public Health Consequences of Sport Specialization," *Journal of Athletic Training* 54, no. 10 (October 1, 2019): 1013–20, https://doi.org/10.4085/1062-6050-521-18.

182 **The repetitive load can add up:** Neeru A. Jayanthi et al., "Sports-Specialized Intensive Training and the Risk of Injury in Young Athletes: A Clinical Case-Control Study," *The American Journal of Sports Medicine* 43, no. 4 (April 2015): 794–801, https://doi.org/10.1177/0363546514567298.

182 **girls may be disproportionately affected:** Neeru A. Jayanthi and Lara R. Dugas, "The Risks of Sports Specialization in the Adolescent Female Athlete," *Strength and Conditioning Journal* 39, no. 2 (April 2017): 20–26, https://doi.org/10.1519/SSC.0000000000000293; Randon Hall et al., "Sport Specialization's Association with an Increased Risk of Developing Anterior Knee Pain in Adolescent Female Athletes," *Journal of Sport Rehabilitation* 24, no. 1 (January 2015): 31–35, https://doi.org/10.1123/jsr.2013-0101.

183 **At the center of this transformation are bones:** Naama W. Constantini and Constance Lebrun, "Bone Health," in *Handbook of Sports Medicine and Science: The Female Athlete*, ed. Margo L. Mountjoy (New Jersey: Wiley, 2015), 57–60.

184 **a "neuromuscular spurt":** Carmen E. Quatman et al., "Maturation Leads to Gender Differences in Landing Force and Vertical Jump Performance: A Longitudinal Study," *The American Journal of Sports Medicine* 34, no. 5 (May 2006): 806–13, https://doi.org/10.1177/0363546505281916.

184 **What's more, young athletes:** Suvi Ravi et al., "Menstrual Dysfunction and Body Weight Dissatisfaction Among Finnish Young Athletes and Non-Athletes," *Scandinavian Journal of Medicine and Science in Sports* 31, no. 2 (February 2021): 405–17, https://doi.org/10.1111/sms.13838; M. L. Rencken, C. H. Chesnut, and B. L. Drinkwater, "Bone Density at Multiple Skeletal Sites in Amenorrheic Athletes," *JAMA* 276, no. 3 (July 17, 1996): 238–40; Kathryn E. Ackerman et al., "Bone Microarchitecture Is Impaired in Adolescent Amenorrheic Athletes Compared with Eumenorrheic Athletes and Nonathletic Controls," *The Journal of Clinical Endocrinology and Metabolism* 96, no. 10 (October 2011): 3123–33, https://doi.org/10.1210/jc.2011-1614.

185 **athletic progression stalls:** Quatman et al., "Maturation Leads to Gender Differences," 806–13.

185 **"perfect storm for significant injuries"**: Andrea Stracciolini et al., "Sex and Growth Effect on Pediatric Hip Injuries Presenting to Sports Medicine Clinic," *Journal of Pediatric Orthopedics B* 25, no. 4 (July 2016): 315–21, https://doi.org/10.1097/BPB.0000000000000315.

186 **awareness of menstrual health:** Kathleen J. Pantano, "Current Knowledge, Perceptions, and Interventions Used by Collegiate Coaches in the U.S. Regarding the Prevention and Treatment of the Female Athlete Triad," *North American Journal of Sports Physical Therapy* 1, no. 4 (November 2006): 195–207; Kathleen J. Pantano, "Knowledge, Attitude, and Skill of High School Coaches with Regard to the Female Athlete Triad," *Journal of Pediatric and Adolescent Gynecology* 30, no. 5 (October 2017): 540–45, https://doi.org/10.1016/j.jpag.2016 .09.013.

186 **Even breast health:** Sarah A. Moore et al., "Exploring the Relationship Between Adolescent Biological Maturation, Physical Activity, and Sedentary Behaviour: A Systematic Review and Narrative Synthesis," *Annals of Human Biology* 47, no. 4 (2020): 365–83, https:// doi.org/10.1080/03014460.2020.1805006.

186 **there's a perception:** Nicole Zarrett, Phillip Veliz, and Don Sabo, *Keeping Girls in the Game: Factors That Influence Sport Participation* (New York: Women's Sports Foundation, 2020).

187 **girls drop out of sports:** Nicole Zarrett and Phillip Veliz, *Teen Sport in America, Part II: Her Participation Matters* (New York: Women's Sports Foundation, 2021), 9–10.

189 **distance runner Alexi Pappas:** Alexi Pappas, *Bravey* (New York: The Dial Press, 2022), 65–66.

190 **Contrary to the dogma:** Arne Güllich, Brooke N. Macnamara, and David Z. Hambrick, "What Makes a Champion? Early Multidisciplinary Practice, Not Early Specialization, Predicts World-Class Performance," *Perspectives on Psychological Science* 17, no. 1 (January 2022): 6–29, https://doi.org/10.1177/1745691620974772.

190 **journalist and author David Epstein:** David Epstein, "My Dad Threw Me into Every Sport You Could Imagine," *Range Widely,* September 21, 2021, https://davidepstein.bulletin .com/2649729328662930/.

191 **Alexi Pappas writes:** Pappas, *Bravey,* 67, 70.

191 **better approach to youth sports:** Gregory D. Myer et al., "Sports Specialization, Part II: Alternative Solutions to Early Sport Specialization in Youth Athletes," *Sports Health* 8, no. 1 (February 2016): 65–73, https://doi.org/10.1177/1941738115614811.

192 **her coach Matt James:** Sean Ingle, "Emma Raducanu Reaps Benefit of Multi-Sports Background," *The Guardian,* September 13, 2021, https://www.theguardian.com/sport/2021/sep/13 /emma-raducanu-reaps-benefit-of-multi-sports-background.

193 **to navigate these waters:** Neeru Jayanthi et al., "Developmental Training Model for the Sport Specialized Youth Athlete: A Dynamic Strategy for Individualizing Load-Response During Maturation," *Sports Health* 14, no. 1 (January 2022): 142–53, https://doi.org/10.1177 /19417381211056088.

195 **gold medalist Jennifer Capriati:** Robin Finn, "The Second Time Around for Jennifer Capriati," *The New York Times,* September 26, 1994, https://www.nytimes.com/1994/09/26 /sports/tennis-the-second-time-around-for-jennifer-capriati.html.

196 **ten-year evaluation:** C. L. Otis et al., "The Sony Ericsson WTA Tour 10 Year Age Eligibility and Professional Development Review," *British Journal of Sports Medicine* 40, no. 5 (May 2006): 464–68, https://doi.org/10.1136/bjsm.2005.023346.

196 **It's not a new concept:** Robert M. Malina et al., "Bio-Banding in Youth Sports: Background, Concept, and Application," *Sports Medicine* 49, no. 11 (November 2019): 1671–85, https://doi.org/10.1007/s40279-019-01166-x.

198 **concludes Tom Hicks:** Marc Serber, "U.S. Soccer Bio-Banding Benefits Coaches in Identifying Talented Players," U.S. Soccer, February 24, 2020, https://www.ussoccer.com/st ories/2020/02/us-soccer-bio-banding-benefits-coaches-in-identifying-talented-players.

199 **open letter and testimonials:** "Women's Cross Country Alumnae Speak Out on Culture of Disordered Eating, Injuries," *Wesleying* (blog), March 2, 2020, http://wesleying.org/2020 /03/02/womens-cross-country-alumnae-speak-out-on-culture-of-disordered-eating-injuries/.

200 **following the Wesleyan letter:** Christine Yu, "Running's Cultural Reckoning Is Long Overdue," *Outside,* May 27, 2021, https://www.outsideonline.com/health/running/running -culture-allegations-wesleyan-university-arizona/.

201 **Ashleigh Barty paused:** Liz Clarke, "Ashleigh Barty Walked Away from Tennis to Find Her Way Forward. Will Other Pros Follow?," *The Washington Post*, May 30, 2021, https://www .washingtonpost.com/sports/2021/05/30/ashleigh-barty-french-open-tennis-career-break/.

202 **announced her retirement:** Ash Barty (@ashbarty), "Today is difficult and filled with emotion for me as I announce my retirement from tennis . . . ," Instagram video, March 22, 2022, https://www.instagram.com/p/Cbbbr7xBX7N/.

10. Family Matters

204 **she had to make a choice:** Aliphine Tuliamuk and Jen Ator, "Aliphine Tuliamuk: What My Dreams Look Like Now," *Women's Running,* April 1, 2020, https://www.womensrun ning.com/culture/people/aliphine-tuliamuk-olympics-postponed-covid-19.

205 **It's an equation that involves:** Dawn Harper-Nelson, "Motherhood and the Olympics," interview by Tina Muir, *Running Realized,* podcast, episode 10, July 12, 2021, https://run ningforreal.com/running-realized-episode-ten/; "Kerri Walsh Jennings: I Was Warned to Not Get Pregnant," interview by Ahiza Garcia, CNN Business, video, February 22, 2018, https:// youtu.be/dSX3T5Bx4mc; Dave Sheinin, Bonnie Berkowitz, and Rick Maese, "They Are Olympians. They Are Mothers. And They No Longer Have to Choose," *The Washington Post*, July 20,2021, https://www.washingtonpost.com/sports/olympics/interactive/2021/olym pics-mothers/.

206 **"kiss of death for a female athlete":** Alysia Montaño, "Nike Told Me to Dream Crazy, Until I Wanted a Baby," *The New York Times,* May 12, 2019, https://www.nytimes.com /2019/05/12/opinion/nike-maternity-leave.html.

207 **maternal and fetal health and safety:** S. Vincent Rajkumar, "Thalidomide: Tragic Past and Promising Future," *Mayo Clinic Proceedings* 79, no. 7 (July 1, 2004): 899–903, https:// doi.org/10.4065/79.7.899; American College of Obstetricians and Gynecologists, "Ethical Considerations for Including Women as Research Participants," Committee Opinion no. 646, 2015, https://www.acog.org/en/clinical/clinical-guidance/committee-opinion/articles/2015 /11/ethical-considerations-for-including-women-as-research-participants.

207 **prenatal exercise programs and guidelines:** Danielle Symons Downs et al., "Physical Activity and Pregnancy: Past and Present Evidence and Future Recommendations," *Research Quarterly for Exercise and Sport* 83, no. 4 (December 2012): 485–502.

208 **opinion of a committee of doctors:** Victoria L. Meah, Gregory A. Davies, and Margie H. Davenport, "Why Can't I Exercise During Pregnancy? Time to Revisit Medical 'Absolute' and 'Relative' Contraindications: Systematic Review of Evidence of Harm and a Call to Action," *British Journal of Sports Medicine* 54, no. 23 (December 2020): 1395–1404, https://doi .org/10.1136/bjsports-2020-102042. Margie Davenport provided the example of the ceiling of 140 beats per minute during an interview.

209 **new Canadian guidelines:** Michelle F. Mottola et al., "2019 Canadian Guideline for Physical Activity Throughout Pregnancy," *British Journal of Sports Medicine* 52, no. 21 (November 2018): 1339–46, https://doi.org/10.1136/bjsports-2018-100056.

210 **It wasn't until 2015:** Kari Bø et al., "Exercise and Pregnancy in Recreational and Elite Athletes: 2016 Evidence Summary from the IOC Expert Group Meeting, Lausanne. Part 1—Exercise in Women Planning Pregnancy and Those Who Are Pregnant," *British Journal of Sports Medicine* 50, no. 10 (May 2016): 571–89, https://doi.org/10.1136/bjsports-2016 -096218; Kari Bø et al., "Exercise and Pregnancy in Recreational and Elite Athletes: 2016 Evidence Summary from the IOC Expert Group Meeting, Lausanne. Part 2—The Effect of Exercise on the Fetus, Labour and Birth," *British Journal of Sports Medicine* 50, no. 21 (November 2016): 1297–1305, https://doi.org/10.1136/bjsports-2016-096810; Kari Bø et al., "Exercise and Pregnancy in Recreational and Elite Athletes: 2016/17 Evidence Summary from the IOC Expert Group Meeting, Lausanne. Part 3—Exercise in the Postpartum Period," *British Journal of Sports Medicine* 51, no. 21 (November 2017): 1516–25, https://doi .org/10.1136/bjsports-2017-097964; Kari Bø et al., "Exercise and Pregnancy in Recreational and Elite Athletes: 2016/17 Evidence Summary from the IOC Expert Group Meeting, Lausanne. Part 4—Recommendations for Future Research," *British Journal of Sports Medicine* 51, no. 24 (December 2017): 1724–26, https://doi.org/10.1136/bjsports-2017-098387; Kari Bø et al., "Exercise and Pregnancy in Recreational and Elite Athletes: 2016/2017 Evidence

Summary from the IOC Expert Group Meeting, Lausanne. Part 5—Recommendations for Health Professionals and Active Women," *British Journal of Sports Medicine* 52, no. 17 (September 2018): 1080–85, https://doi.org/10.1136/bjsports-2018-099351.

211 **athletes don't experience more complicated deliveries:** Thorgerdur Sigurdardottir et al., "Do Female Elite Athletes Experience More Complicated Childbirth than Non-Athletes? A Case-Control Study," *British Journal of Sports Medicine* 53, no. 6 (March 2019): 354–58, https://doi.org/10.1136/bjsports-2018-099447.

214 **trauma from pregnancy and childbirth:** Malin Huber, Ellen Malers, and Katarina Tunón, "Pelvic Floor Dysfunction One Year After First Childbirth in Relation to Perineal Tear Severity," *Scientific Reports* 11, no. 1 (June 15, 2021): 12560, https://doi.org/10.1038/s41598-021-91799-8.

214 **rock climber Beth Rodden:** Megan Michelson, "How Elite Athletes Come Back After Childbirth," *Outside*, March 28, 2018, https://www.outsideonline.com/health/training-performance/how-elite-athletes-come-back-after-childbirth/.

215 **ramifications for exercise and performance:** Isabel S. Moore et al., "Multidisciplinary, Biopsychosocial Factors Contributing to Return to Running and Running Related Stress Urinary Incontinence in Postpartum Women," *British Journal of Sports Medicine* 55, no. 22 (November 2021): 1286–92, https://doi.org/10.1136/bjsports-2021-104168.

215 **more than a cosmetic issue:** Rita E. Deering et al., "Impaired Trunk Flexor Strength, Fatigability, and Steadiness in Postpartum Women," *Medicine and Science in Sports and Exercise* 50, no. 8 (August 2018): 1558–69, https://doi.org/10.1249/MSS.0000000000001609; Rita E. Deering et al., "Fatigability of the Lumbopelvic Stabilizing Muscles in Women 8 and 26 Weeks Postpartum," *Journal of Women's Health Physical Therapy* 42, no. 3 (2018): 128–38, https://doi.org/10.1097/JWH.0000000000000109.

216 **mom of three Alysia Montaño:** Alysia Montaño (@alysiamontano), "Ok sooooo . . . here are my photos from 1 week post op . . . ," Instagram photo, December 7, 2020, https://www.instagram.com/p/CIhWVvzj7Zd/.

216 **fall into an energy debt:** Gráinne Donnelly, Emma Brockwell, and Tom Goom, "Ready, Steady . . . GO! Ensuring Postnatal Women are Run-Ready!," *Blog, British Journal of Sports Medicine*, May 20, 2019, https://blogs.bmj.com/bjsm/2019/05/20/ready-steadygo-ensuring-postnatal-women-are-run-ready/.

216 **breast milk production:** Kari Bø et al., "Exercise and Pregnancy in Recreational and Elite Athletes. Part 3," 1516–25.

216 **breastfeeding her daughter, Zoe:** Aliphine Tuliamuk (@aliphine), "Motherhood part 4: BREASTFEEDING . . . ," Instagram photo, June 6, 2021, https://www.instagram.com/p/CPyk5j2nOQT/.

218 **do some Kegels:** Padma Kandadai, Katharine O'Dell, and Jyot Saini, "Correct Performance of Pelvic Muscle Exercises in Women Reporting Prior Knowledge," *Female Pelvic Medicine and Reconstructive Surgery* 21, no. 3 (June 2015): 135–40, https://doi.org/10.1097/SPV.0000000000000145.

220 **study with recreational runners:** Rita E. Deering et al., "Exercise Program Reduces Inter-Recti Distance in Female Runners Up to 2 Years Postpartum," *Journal of Women's Health Physical Therapy* 44, no. 1 (March 2020): 9–18, https://doi.org/10.1097/JWH.0000000000000157.

221 **On Mother's Day 2019:** Montaño, "Nike Told Me to Dream Crazy."

222 **Felix's contract with Nike:** David Marchese, "Allyson Felix Knows What Really Makes the Olympics Run," *The New York Times*, June 16, 2021, https://www.nytimes.com/interactive/2021/06/14/magazine/allyson-felix-interview.html; Allyson Felix, "Allyson Felix: My Own Nike Pregnancy Story," *The New York Times*, May 22, 2019, https://www.nytimes.com/2019/05/22/opinion/allyson-felix-pregnancy-nike.html; Sean Gregory, "Motherhood Could Have Cost Olympian Allyson Felix. She Wouldn't Let It," *Time*, July 8, 2021, https://time.com/6077124/allyson-felix-tokyo-olympics/.

224 **"a new paradigm for postpartum care":** American College of Obstetricians and Gynecologists, "Optimizing Postpartum Care," Committee Opinion no. 736, May 2018, https://www.acog.org/clinical/clinical-guidance/committee-opinion/articles/2018/05/optimizing-postpartum-care.

224 ***Returning to Running Postnatal:*** G. Donnelly, E. Brockwell, and T. Goom, "Return to Running Postnatal—Guideline for Medical, Health and Fitness Professionals Managing This Population," *Physiotherapy* 107, Suppl. 1 (May 1, 2020): e188–89, https://doi.org/10 .1016/j.physio.2020.03.276.

225 **current system of postpartum care:** Rita E. Deering, Shefali M. Christopher, and Bryan C. Heiderscheit, "From Childbirth to the Starting Blocks: Are We Providing the Best Care to Our Postpartum Athletes?," *Journal of Orthopaedic and Sports Physical Therapy* 50, no. 6 (June 2020): 281–84, https://doi.org/10.2519/jospt.2020.0607.

225 **Elana Meyers Taylor:** Emily Kraus, et al., "Bridging the Research-Practice Gap in Women's Sport," session at the 2021 Female Athlete Conference, virtual, June 10, 2021.

11. The Change

230 **menopause is simply another life stage:** Jen Gunter, *The Menopause Manifesto* (New York: Citadel Press, 2021), 7, 26–37, 55–59; Kimberly Peacock and Kari M. Ketvertis, "Menopause," in *StatPearls* (Treasure Island, FL: StatPearls Publishing, 2022), http://www .ncbi.nlm.nih.gov/books/NBK507826/.

231 **a long list of symptoms:** Nanette Santoro, "Perimenopause: From Research to Practice," *Journal of Women's Health* 25, no. 4 (April 2016): 332–39, https://doi.org/10.1089/jwh .2015.5556.

231 **Study of Women's Health Across the Nation:** Ellen B. Gold et al., "Factors Related to Age at Natural Menopause: Longitudinal Analyses from SWAN," *American Journal of Epidemiology* 178, no. 1 (July 1, 2013): 70–83, https://doi.org/10.1093/aje/kws421.

232 **the deterioration of physical health:** Susan Mattern, "What if We Didn't Dread Menopause?," *The New York Times*, September 12, 2019, https://www.nytimes.com/2019/09/12 /opinion/sunday/menopause-symptoms.html; Jessica Grose, "Why Is Perimenopause Still Such a Mystery?," *The New York Times*, April 29, 2021, https://www.nytimes.com/2021/04 /29/well/perimenopause-women.html.

233 **menopause became a disease:** Grace E. Kohn et al., "The History of Estrogen Therapy," *Sexual Medicine Reviews* 7, no. 3 (July 2019): 416–21, https://doi.org/10.1016/j.sxmr.2019 .03.006.

233 **a July 2002 study:** Writing Group for the Women's Health Initiative Investigators, "Risks and Benefits of Estrogen Plus Progestin in Healthy Postmenopausal Women: Principal Results from the Women's Health Initiative Randomized Controlled Trial," *JAMA* 288, no. 3 (July 17, 2002): 321–33, https://doi.org/10.1001/jama.288.3.321. A UK study found a similar link: "Breast Cancer and Hormone-Replacement Therapy in the Million Women Study," *The Lancet* 362, no. 9382 (August 9, 2003): 419–27, https://doi.org/10.1016/S0140-6736(03) 14065-2.

233 **"This is the biggest bombshell":** Gina Kolata and Melody Petersen, "Hormone Replacement Study A Shock to the Medical System," *The New York Times*, July 10, 2002, https:// www.nytimes.com/2002/07/10/us/hormone-replacement-study-a-shock-to-the-medical -system.html.

234 *JAMA* **study was widely criticized:** Tara Parker-Pope, "How NIH Misread Hormone Study in 2002," *The Wall Street Journal*, July 9, 2007, https://www.wsj.com/articles /SB118394176612760522; John A Goldman, "The Women's Health Initiative 2004 - Review and Critique," *Medscape General Medicine* 6, no. 3 (August 9, 2004): 65, https://www.ncbi .nlm.nih.gov/pmc/articles/PMC1435607/.

234 **Long-term follow-up studies:** JoAnn E. Manson et al., "Menopausal Hormone Therapy and Long-Term All-Cause and Cause-Specific Mortality: The Women's Health Initiative Randomized Trials," *JAMA* 318, no. 10 (September 12, 2017): 927–38, https://doi.org /10.1001/jama.2017.11217; Rowan T. Chlebowski et al., "Association of Menopausal Hormone Therapy with Breast Cancer Incidence and Mortality During Long-Term Follow-Up of the Women's Health Initiative Randomized Clinical Trials," *JAMA* 324, no. 4 (July 28, 2020): 369–80, https://doi.org/10.1001/jama.2020.9482.

235 **According to a 2019 survey:** Juliana M. Kling et al., "Menopause Management Knowledge in Postgraduate Family Medicine, Internal Medicine, and Obstetrics and Gynecology

Residents: A Cross-Sectional Survey," *Mayo Clinic Proceedings* 94, no. 2 (February 2019): 242–53, https://doi.org/10.1016/j.mayocp.2018.08.033.

235 **seek out specialty training:** For example, NAMS certifies clinicians as menopause practitioners. See "NCMP Certification," The North American Menopause Society, https://www.menopause.org/for-professionals/ncmp-certification.

235 **estimated 1.3 million menstruating people:** Peacock and Ketvertis, "Menopause."

235 **Amanda Thebe was forty-two:** Sources include an interview with Amanda Thebe; Amanda Thebe, *Menopocalypse: How I Learned to Thrive During Menopause and How You Can Too* (Vancouver/Berkeley: Greystone Books, 2020), 1–6.

238 **"You might try to push harder":** Lindsay Warner, "The Athlete's Guide to Menopause," *Outside*, December 8, 2020, https://www.outsideonline.com/health/training-performance/menopause-exercise-tips/.

239 **Aerobic fitness decreases with age:** Amanda Q. X. Nio et al., "The Menopause Alters Aerobic Adaptations to High-Intensity Interval Training," *Medicine and Science in Sports and Exercise* 52, no. 10 (October 2020): 2096–106, https://doi.org/10.1249/MSS.0000000000002372.

239 **Body composition and proportions morph:** Lacey M. Gould et al., "Metabolic Effects of Menopause: A Cross-Sectional Characterization of Body Composition and Exercise Metabolism," *Menopause* (February 28, 2022), https://doi.org/10.1097/GME.0000000000001932; Gail A. Greendale et al., "Changes in Body Composition and Weight During the Menopause Transition," *JCI Insight* 4, no. 5 (March 7, 2019), https://doi.org/10.1172/jci.insight.124865; J. C. Lovejoy et al., "Increased Visceral Fat and Decreased Energy Expenditure During the Menopausal Transition," *International Journal of Obesity* 32, no. 6 (June 2008): 949–58, https://doi.org/10.1038/ijo.2008.25.

240 **the accumulating fat:** Julie Abildgaard et al., "Changes in Abdominal Subcutaneous Adipose Tissue Phenotype Following Menopause Is Associated with Increased Visceral Fat Mass," *Scientific Reports* 11, no. 1 (July 20, 2021): 14750, https://doi.org/10.1038/s41598-021-94189-2.

240 **consequence of tanking hormones:** Mette Hansen and Michael Kjaer, "Influence of Sex and Estrogen on Musculotendinous Protein Turnover at Rest and After Exercise," *Exercise and Sport Sciences Reviews* 42, no. 4 (October 2014): 183–92, https://doi.org/10.1249/JES.0000000000000026; Nkechinyere Chidi-Ogbolu and Keith Baar, "Effect of Estrogen on Musculoskeletal Performance and Injury Risk," *Frontiers in Physiology* 9 (2019), https://doi.org/10.3389/fphys.2018.01834.

241 **researchers followed women in Finland:** Dmitriy Bondarev et al., "Physical Performance in Relation to Menopause Status and Physical Activity," *Menopause* 25, no. 12 (December 2018): 1432–41, https://doi.org/10.1097/GME.0000000000001137.

243 **Add in sleep troubles:** Howard M. Kravitz and Hadine Joffe, "Sleep During the Perimenopause: A SWAN Story," *Obstetrics and Gynecology Clinics of North America* 38, no. 3 (September 2011): 567–86, https://doi.org/10.1016/j.ogc.2011.06.002.

244 **Research indicates that HIIT:** Marine Dupuit et al., "Effect of High Intensity Interval Training on Body Composition in Women Before and After Menopause: A Meta-Analysis," *Experimental Physiology* 105, no. 9 (2020): 1470–90, https://doi.org/10.1113/EP088654.

244 **combination of hormone therapy and strength training:** Paula H. A. Ronkainen et al., "Postmenopausal Hormone Replacement Therapy Modifies Skeletal Muscle Composition and Function: A Study with Monozygotic Twin Pairs," *Journal of Applied Physiology* 107, no. 1 (July 2009): 25–33, https://doi.org/10.1152/japplphysiol.91518.2008; Sarianna Sipilä et al., "Effects of Hormone Replacement Therapy and High-Impact Physical Exercise on Skeletal Muscle in Post-Menopausal Women: A Randomized Placebo-Controlled Study," *Clinical Science* 101, no. 2 (August 2001): 147–57, https://pubmed.ncbi.nlm.nih.gov/11473488/.

244 **guard against tendon injuries:** Chidi-Ogbolu and Baar, "Effect of Estrogen on Musculoskeletal Performance and Injury Risk."

245 **ultrarunner Magda Boulet:** Liza Howard, "Age-Old Runners: Magda Boulet," iRunFar, July 1, 2020, https://www.irunfar.com/age-old-runners-magda-boulet.

247 **a $600 billion market:** Emma Hinchliffe, "Menopause Is a $600 Billion Opportunity, Report Finds," *Fortune*, October 26, 2020, https://fortune.com/2020/10/26/menopause-startups-female-founders-fund-report/.

Conclusion: Beyond the Gap

250 **an image circulated on social media:** Ali Kershner (@kershner.ali), "Not usually one for this type of post . . . ," Instagram photo, March 18, 2021, https://www.instagram.com/p/CMkRJ2LswFp/.

250 **Sedona Prince posted a video:** Sedona Prince (@sedonerrr), "it's 2021 and we are still fighting for bits and pieces of equality. #ncaa #inequality #fightforchange," TikTok, March 18, 2021, https://www.tiktok.com/@sedonerrr/video/6941180880127888646?lang=en.

251 **it wasn't out of the ordinary:** Lindsay Gibbs, "Unlocking the NCAA Gender Inequity Files," *Power Plays*, March 27, 2022, https://www.powerplays.news/p/unlocking-the-ncaa-gender-inequity?s=r; Henry Bushnell, "NCAA Reveals Budget, Revenue Gulfs Between Men's and Women's Basketball Tournaments," *Yahoo! Sports*, March 26, 2021, https://www.yahoo.com/lifestyle/ncaa-revenue-budget-march-madness-mens-womens-basketball-tournaments-202859329.html.

251 **Coaches, WNBA players, and others:** Cat Ariail, "NCAAW Tournament: Women's Basketball World Holds NCAA Accountable for Inequities," *Swish Appeal*, SBNation, March 23, 2021, https://www.swishappeal.com/ncaa/2021/3/23/22344510/ncaaw-tournament-gender-inequities-disparities-weight-room-food-swag-sedona-prince-dawn-staley; Muffet McGraw (@MuffetMcGraw), photo, Twitter, March 20, 2021, https://twitter.com/MuffetMcGraw/status/1373321930485473287.

252 **the first of two reports:** Kaplan Hecker & Fink LLP, "NCAA External Gender Equity Review, Phase I: Basketball Championships," August 2, 2021, https://ncaagenderequityreview.com/.

253 **As Anthony Weems:** Brendon Kleen, "A New Case for Diversity in College Sports," *Global Sports Matters*, May 7, 2021, https://globalsportsmatters.com/business/2021/05/07/new-case-diversity-college-sports-ncaa/.

255 **an average of seventeen years:** Alice Ammerman, Tosha Woods Smith, and Larissa Calancie, "Practice-Based Evidence in Public Health: Improving Reach, Relevance, and Results," *Annual Review of Public Health* 35 (2014): 47–63, https://doi.org/10.1146/annurev-publhealth-032013-182458.

258 **WNBA player Layshia Clarendon:** Frankie de la Cretaz, "Living Nonbinary in a Binary Sports World," *Sports Illustrated*, April 16, 2021, https://www.si.com/wnba/2021/04/16/nonbinary-athletes-transgender-layshia-clarendon-quinn-rach-mcbride-daily-cover.

259 **On January 16 in Houston, Texas:** Taylor Dutch, "Two American Records Fall at the Houston Marathon and Half Marathon," *Runner's World*, January 16, 2022, https://www.runnersworld.com/news/a38784069/houston-marathon-results-keira-damato-sara-hall/.

260 **She also tweeted:** Keira D'Amato (@KeiraDAmato), "After setting the American Record in the Marathon . . . ," Twitter, January 17, 2022, https://twitter.com/KeiraDAmato/status/1483085297055375364.

260 **It was a historic moment:** Jake Howard, "Day 2 Recap: History Made as First Women's Championship Tour Event at Pipe Underway," *World Surf League*, January 30, 2022, https://www.worldsurfleague.com/posts/495688/2022-billabong-pro-pipeline-coverage.

Index